THE 4TH DIMENSION AND BEYOND

Imagining Worlds with 0, 1, 2, 3, 4 Dimensions and More

John Eikum

Beaver's Pond Press, Inc.

Edina, Minnesota

THE 4TH DIMENSION AND BEYOND: IMAGINING WORLDS WITH 0, 1, 2, 3, 4 DIMENSIONS AND MORE
© copyright 2007 by John Eikum. All rights reserved. No part of this book may be reproduced in any form whatsoever, by photography or xerography or by any other means, by broadcast or transmission, by translation into any kind of language, nor by recording electronically or otherwise, without permission in writing from the author, except by a reviewer, who may quote brief passages in critical articles or reviews.

ISBN 10: 1-59298-172-0
ISBN 13: 978-1-59298-172-4

Library of Congress Control Number: 2006939102
Printed in the United States of America
First Printing: March 2007
11 10 09 08 07   6 5 4 3 2 1

Cover and interior design by Clay Schotzko
Edited by Arlene Prunkl

Beaver's Pond Press, Inc.

Beaver's Pond Press, Inc.
7104 Ohms Lane, Suite 216
Edina, MN 55439-2129
(952) 829-8818
www.BeaversPondPress.com
Beaver's Pond Press is an imprint of the Adams Publishing Group.

To order, visit www.BookHouseFulfillment.com or call 1-800-901-3480. Reseller discounts available.

Dedication

To Megan, Andy, Carrie,

and inquisitive children throughout the world.

Exercise your imagination.

Go after your dreams.

TABLE OF CONTENTS

1. **WHAT ARE DIMENSIONS?** 7
2. **A VISIT TO A WORLD CALLED FLATLAND** 17
3. **WHY CAN'T A FLATLANDER IMAGINE SPACELAND?** 31
   - ANALOGY 1: RIGHT NEXT TO YOU 32
   - ANALOGY 2: SEEING ENTIRE SHAPES 33
   - ANALOGY 3: SEEING INSIDE 34
   - ANALOGY 4: MAKING THINGS DISAPPEAR.......36
   - ANALOGY 5: CROSS SECTIONS 36
4. **DIMENSIONS AND SHAPES** 39
5. **A 4-DIMENSIONAL CUBE** 47
6. **WHAT IS 4D-LAND LIKE?** 59
   - ANALOGY 1: RIGHT NEXT TO YOU 61
   - ANALOGY 2: SEEING ENTIRE SHAPES 61
   - ANALOGY 3: SEEING INSIDE 62
   - ANALOGY 4: MAKING THINGS DISAPPEAR.......63
   - ANALOGY 5: CROSS SECTIONS 64
   - ANALOGY 6: ROTATION 64
   - ANALOGY 7: MAKING SPHERES 66
7. **HOW MANY PARTS DOES A CUBE HAVE?** 71
8. **HOW CAN WE TRY TO IMAGINE 4D-LAND?**....81
   - SHOW ME THE NEXT DIMENSION 81
   - LOOKING AT CUBES 84
   - HOW TO FOLD A CUBE 89
9. **ISN'T THE 4TH DIMENSION TIME?** 97
10. **IS THIS USEFUL?** 99
    - HYPERCUBE COMPUTER 100
    - DATA ANALYSIS 103
    - UNDERSTANDING OUR UNIVERSE 105

   THE HISTORY OF FLATLAND 109

   INDEX 111

WHAT ARE DIMENSIONS?

Have you ever thought about the different directions in which you can move? Moving around is something we usually don't think much about—we just do it naturally. By looking more closely, however, at the directions we can move in, we can better understand the idea of **dimensions.**

To learn about dimensions, let's look at a checkerboard. We won't be concerned about the rules for the game of checkers; we're going to use the checkerboard pattern to discuss motion, direction, and dimensions.

Place a checker on a square near the middle of a checkerboard. You can move the checker forward and backward to squares in front of and behind it. Notice that these two directions are on the same line. If you move the checker forward three squares and then backward three squares, it ends up on the same square that it

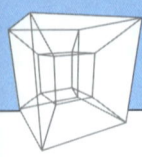

The Fourth Dimension and Beyond

started on. Let's call the pair of directions on this line **forward/backward**.

You can also move the checker on a line of squares to the left and right. Let's call the pair of directions on this line **left/right**.

Using these two pairs of directions, forward/backward and left/right, you can move the checker from any square on the board to any other square. In the following picture, the checker is moved right two squares

What Are Dimensions?

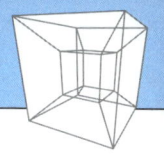

and forward three squares.

You could make a huge checkerboard with millions of squares and you would still be able to move from any square to any other square by moving forward/backward and left/right. With *only* forward/backward or *only* left/right, however, it would be impossible to move to every square. You need two pairs of directions to move from any square to any other square, and you never need more than that.

You could, of course, also move the checker diagonally through the corners of the squares. This is a shortcut but would require an infinite number of directions and doesn't allow moving to any additional places. Only forward/backward and left/right, in the correct combination, are required to move from any square to any other square.

Each of these pairs of directions represents a

The Fourth Dimension and Beyond

dimension and since two pairs of directions are needed to move on a checkerboard, it is said to have two dimensions. You can also say that the board is **2-dimensional**. One dimension is the forward/backward direction and the other is the left/right direction.

You have to be able to move in two dimensions to be able to move from any square to all other squares on a checkerboard. One dimension is not enough, but you never need more than two.

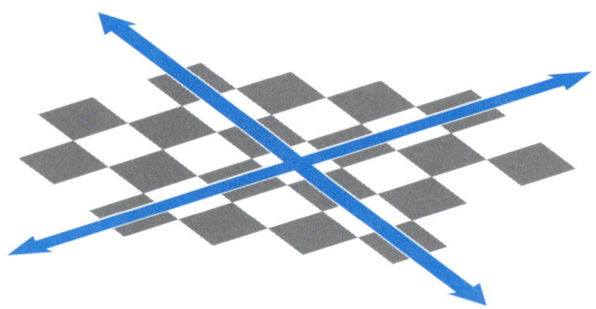

Dimensions, by the way, are not limited in size. Just imagine that the checkerboard pattern extends in every direction forever.

Now look at the corners of the squares. Each of the corners in every square of a checkerboard is a **right angle**. The corners of the pages in this book are also

What Are Dimensions?

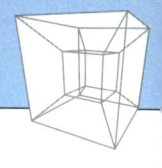

right angles. In fact, right angles are very common; if you look around, you will find many right angles.

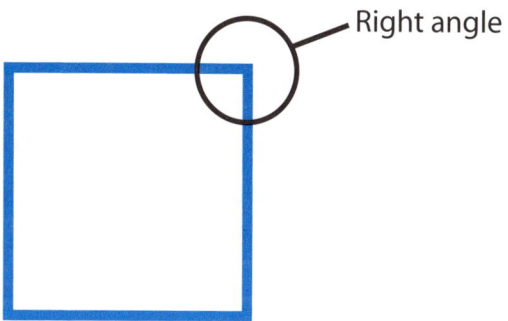

When two lines cross with right angles, they are said to be **perpendicular**.

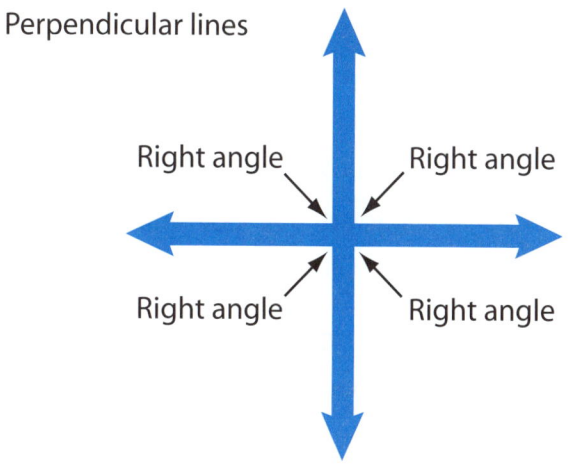

Perpendicular lines

11

The Fourth Dimension and Beyond

The two dimensions that we described on the checkerboard cross at right angles, so they are perpendicular. This idea of perpendicular dimensions will be important as we learn about worlds with more dimensions.

Now, if you think about the world we live in, you will realize there are places that are not on the checkerboard, no matter how large we make it. To see this, pick up the checker, lifting it up above the checkerboard.

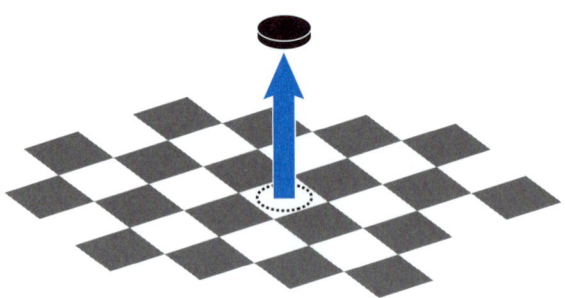

There is no way to move the checker to this position using only the forward/backward and left/right directions on the board.

To move the checker to this position, we need to add a new pair of directions. Let's call this pair **up/down**. Up refers to lifting the checker above the board and down would be moving the checker below the board

What Are Dimensions?

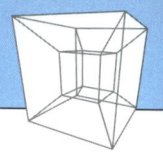

(for example, down to the floor below the table that the checkerboard is on).

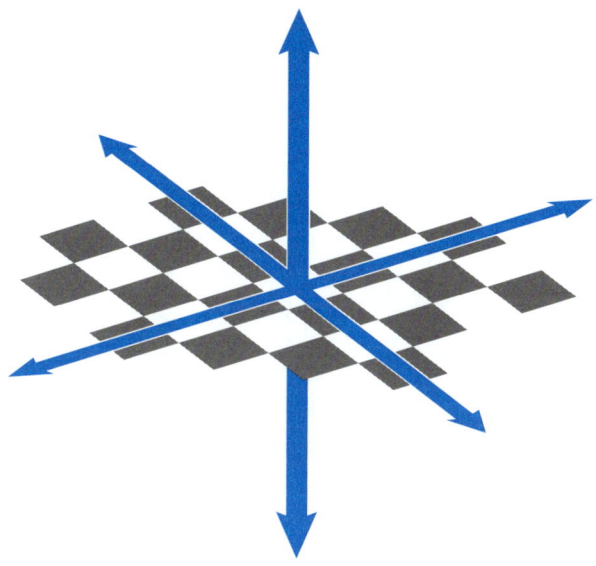

You can move from any place to any other place by using just these three pairs of directions, forward/backward, left/right, and up/down.

It is possible to go straight from each place to every other place, but think about how many directions that would be! In real life, of course, we usually take the shortest, most direct way to get from one place to another. The purpose of learning about dimensions isn't

The Fourth Dimension and Beyond

to make you take the long way to get somewhere. Perpendicular dimensions are useful because they provide a precise way to describe the space we live in.

Now, put your finger on top of the checker so your finger goes straight up from the checkerboard.

Your finger is perpendicular to each of the two dimensions of the checkerboard. And the dimensions of the checkerboard are perpendicular to your finger. Your finger is the third dimension, the up/down dimension.

We've discovered that we live in a **3-dimensional** world. Three (and never more than three) pairs of perpendicular directions are needed to be able to move to all parts of our world. Because each of the dimensions is perpendicular to each of the others, the three dimensions are said to be **mutually perpendicular**.

What Are Dimensions?

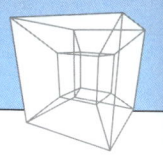

Many of the things in our world are also 3-dimensional. For example, most boxes are made with sides that are rectangles, and the edges are perpendicular to one another.

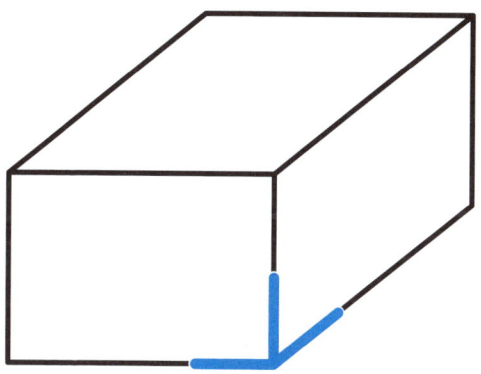

If you look at any corner of such a box, you can see that the edges, meeting at the corner, show the three dimensions. The edges are mutually perpendicular. Our imaginary checkerboard pattern is infinitely large; it extends forever. In our 3-dimensional world, the three perpendicular dimensions also extend forever. An actual object, like a box, is not infinitely large and its size in each of the dimensions can be measured. These sizes are often referred to as the length, width, and height of the box.

The Fourth Dimension and Beyond

Many shapes don't have straight edges that meet at corners. A circle, for example, has no straight parts, but it is 2-dimensional. To see this, notice that you can draw a circle on a checkerboard. You need both the forward/backward and the left/right directions of the checkerboard to draw a circle, but you don't need to leave the checkerboard by moving in the up/down direction.

A round ball is called a **sphere**. Although a sphere doesn't have straight edges like a box, it is a 3-dimensional shape. The forward/backward and the left/right directions of the checkerboard are not enough to create a sphere; you also need the up/down direction.

In this book, we will learn about and compare imaginary worlds that have zero, one, two, four, and more dimensions. And along the way we'll compare them to our own world of three dimensions.

A VISIT TO A WORLD CALLED FLATLAND

I'd like to tell you a story about an imaginary place called Flatland.

Like the checkerboard, Flatland is a 2-dimensional world. For example, the top of your table could be Flatland. The inhabitants of Flatland are also 2-dimensional and are called Flatlanders. Flatlanders are shapes, like squares, circles, and triangles.

This picture shows a square that lives in a Flatland house. Remember that this is a 2-dimensional house. In this story, it has no height above the paper.

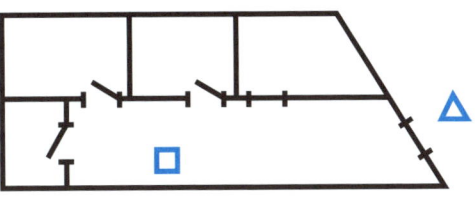

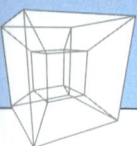

The Fourth Dimension and Beyond

The square's house has several rooms. As you can see, there is a door for each room and also a door to the outside. If the square's outside door is shut, other Flatlanders cannot enter or see into the square's house. In the picture, the triangle can travel around the square's house, but it cannot go inside unless the square opens the door.

Imagine what the house would look like if you were the triangle. Leaving it open, lay this book flat on a table and move your head down toward the edge of the table. Look across the page at the picture of the Flatland house. By doing this, you are almost entering Flatland.

These pictures show how the Flatland house looks as your head moves closer to the table:

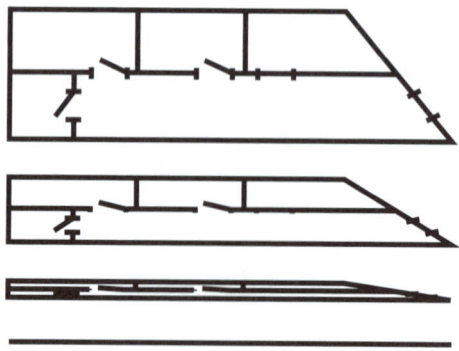

18

A Visit to a World Called Flatland

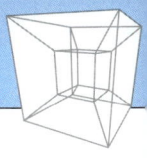

When the triangle looks at the house, it sees only a line because Flatlanders are not able to rise up above the paper like we can to look down into the house.

When Flatlanders look at each other, they also see lines. They learn about each other by counting the number of sides they each have and by determining how long their sides are. Some are triangles. Some are squares. Some Flatlanders have many sides. The circles don't have any straight sides. All of the Flatlanders are flat; they are 2-dimensional.

One day, as the square was having breakfast, it heard someone say, "Good morning." The square was startled and searched its house trying to find who was there. It searched every room but could not find anyone in the house. It checked the door to the outside and found that it was closed.

The puzzled square then heard the strange voice again. "I said, 'Good morning.' Aren't you courteous enough to reply?"

"Who are you and where are you hiding?" asked the square.

"I am a sphere and I am not hiding. I am above you,"

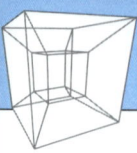

The Fourth Dimension and Beyond

replied the voice. "I am a 3-dimensional being from a place called Spaceland. I have come to visit Flatland."

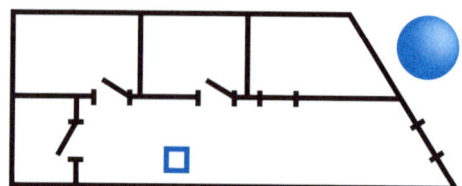

The square was frightened. "I do not know what a 'sphere' is, and I do not understand what you mean when you say 'above.' I have never heard those words before. Now please let me know where you are so I can see you."

"I am looking inside your house and I can see you in your living room."

"That is impossible. No one can see inside my house unless they are in the house with me. My door is shut. I want to know how you got into my house and where you are."

The sphere laughed. "I told you I am not in your house, I am above it. I can see inside your house even though I am not in your house with you. Here, let me show you something."

A Visit to a World Called Flatland

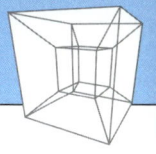

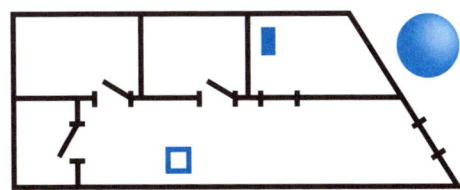

"There is a book in your bedroom and your bedroom door is shut. Right?"

"Yes."

"I am going to pick up the book and put it in your living room without opening your bedroom door."

Suddenly, the square saw the book appear in the living room. It didn't move into the room; it was just suddenly there.

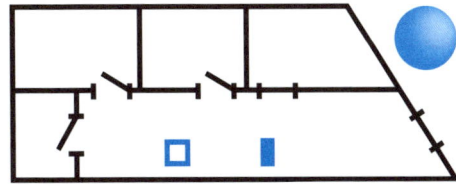

"How did you do that?" asked the square, who was becoming very interested in this strange visitor.

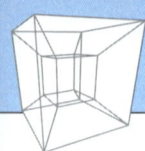

The Fourth Dimension and Beyond

"I am a 3-dimensional being and you are 2-dimensional," replied the sphere. "You can move forward and backward. You can also move left and right. I can move like that too, but I can also move up and down. I picked up your book, lifted it into the 3rd dimension over the bedroom wall, and set it down again.

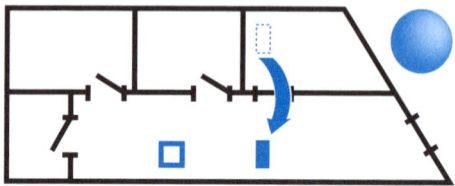

"To move up or down, all you have to do is move in a direction that is perpendicular to the two pairs of directions that you already know."

Now the square was completely confused. "That is crazy," it said. "I understand forward and backward and I understand left and right. I know that those directions are perpendicular to each other, but there is no direction that is perpendicular to both of them. That is just impossible."

"Yes, I thought you would think so," replied the sphere. "It is very difficult for you to understand three dimen-

A Visit to a World Called Flatland

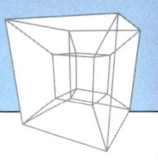

sions when you live in a 2-dimensional world. Here, let me show you what I look like."

"How are you going to do that? I can't even see you."

"I will move into your world. Because you are 2-dimensional, you will not be able to see all of me at once. But you will be able to see parts of me. Are you ready?"

The square wasn't sure it wanted to be visited by a being that it could hear but not see. It had never even heard of three dimensions and now a 3-dimensional thing wanted to come into its house. But the square was curious and wanted to learn more, so it agreed.

Suddenly a shape appeared in the square's living room.

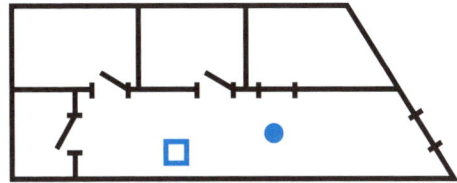

The square investigated the shape and discovered that it was just a small circle. It felt rather relieved and said, "I still don't know how you got into my house, but I

23

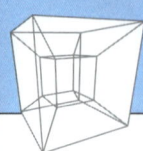

The Fourth Dimension and Beyond

looked at you and now I know that you are just an ordinary circle." As soon as the square said this, the circle began to get larger. This was something the square had never seen before; the creatures of Flatland don't change size like that.

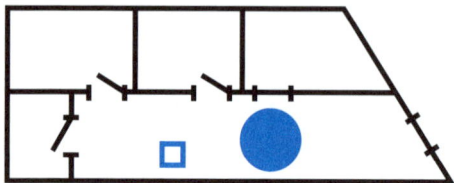

Now the square knew that there was something unusual about this thing that called itself a "sphere," but it still didn't understand three dimensions. So it said, "Okay, I understand. A sphere is a circle that can change its size."

The sphere laughed again and said, "That is close, but actually I am many different-sized circles all at the same time. Try this. Picture a series of circles, starting with one that looks like just a dot and getting larger in size up to the size that you see when you look at me now. Next, make another set that decreases in size until the smallest one is a dot.

A Visit to a World Called Flatland

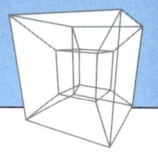

"Now comes the hard part. I know this won't make sense to you, but try to imagine stacking these disks up in order from the smallest to the largest and back down to the smallest. Don't put them in a row, but lift each one up and place it on top of the one before it."

At this point, the square yelled, "Stop! I do not understand this at all. I can put the circles in a line in order by size as you described, but I do not understand what you mean about putting a circle on top of another circle. I just don't know how to do that. I do not understand 'on top.'"

The sphere said, "Well, I can put them on top for you. Oh! Now I see the problem. If I do that, you won't be able to see them. You will just see them disappear as I pick each one up. Since you live in a 2-dimensional world, there is no way you can move things from your world into the 3rd dimension, and you cannot see things that exist outside of your world."

The square was now very frustrated. It had learned about a new world with more dimensions but was not able to understand it. The square was tired and decided to go to bed and get some sleep. It thought, "Maybe I'll be able to understand this tomorrow."

The Fourth Dimension and Beyond

That night, as the square slept, it had a dream. In its dream, it visited a world called Lineland. In Lineland, all of the inhabitants are straight lines. Some are short, some are long, and others are in between. All of the Linelanders are lined up end to end in a straight line.

Lineland

The Linelanders can move back and forth in the line but can't move past their neighbors, and they can't move out of the line. Lineland is 1-dimensional.

In its dream, the square moved close to one of the Linelanders and spoke to it.

The Linelander was shocked. The voice didn't come from either of its neighbors. In fact, it seemed to the Linelander that the voice came from inside itself.

The square said, "Hello. Will you talk to me?"

The Linelander replied, "I don't know. I can't figure out where you are."

A Visit to a World Called Flatland

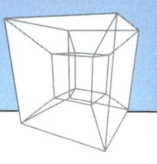

"I'm right here beside you."

"No, my two neighbors are beside me. I can see them both; they look like points. I know that you are not one of my neighbors."

"Well," said the square, "all you have to do is look to the side and you can see me."

"I have no idea what you are talking about. I don't understand 'look to the side.'"

"I am at your side and I can see you: both ends of you as well as your middle. And I can see both of your neighbors and their neighbors and so on way down the line."

"See my middle? And both of my ends at the same time? This is crazy. Do you expect me to believe any of this?"

Then the square had an idea. "Since you can only see your two neighbors, I'll move into the space between you and one of your neighbors. Then you can see me."

"I want to know what you are!" cried the Linelander in frustration.

The Fourth Dimension and Beyond

"I am a square and I have come from Flatland to visit you."

"I've never heard of a square. What is that?"

"I am made of four lines…"

"Wait a minute!" the Linelander interrupted. "Everyone in Lineland has two ends connected by one line. Four lines? That just doesn't make any sense."

At that moment, the square moved into the space between the Linelander and its neighbor so that it crossed Lineland. "There, now you can see me."

The Linelander said, "you look like a point, just like everyone else in Lineland."

"I am going to show you how to move sideways. I'm moving right now."

The Linelander said, "I still see you and you are still a point."

A Visit to a World Called Flatland

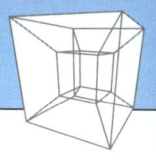

The square kept moving sideways until it moved out of Lineland. The Linelander screamed, "You just disappeared. Where did you go? How can you disappear?"

At this point, the square decided that it was hopeless to try to explain two dimensions to a line that only knows about one dimension.

Later in its dream, the square visited a tiny dot. The dot was unable to move in any direction. A dot that is really, really tiny is called a **point**. In fact, a point is so small that it has no dimensions. The point lived in Pointland, and Pointland is 0-dimensional.

Pointland •

The square tried to talk to the point, but the point didn't even answer. It just sat in Pointland all alone.

When the square awoke the next morning, the circle that called itself a sphere was still in the square's living room. The square told the sphere about its dream.

The sphere listened intently to everything the square had to say. As it listened to the square explain how frustrating it was to try to explain two dimensions to the Linelander, the sphere suddenly realized why it is

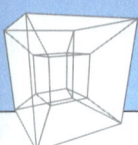

The Fourth Dimension and Beyond

so difficult to explain three dimensions to the square.

It is easy to understand fewer dimensions—the sphere understood Flatland, Lineland, and Pointland, and the square understood Lineland and Pointland. But it is difficult to understand more dimensions—the Linelander couldn't imagine Flatland and neither the Linelander nor the Flatlander was able to understand Spaceland.

WHY CAN'T A FLATLANDER IMAGINE SPACELAND?

When the sphere heard the story from the square about trying to explain two dimensions to the Linelander, it understood why it is so difficult to explain three dimensions to the Flatlander.

An **analogy** helps us understand something by comparing it to something similar. Analogies are a good way to understand things that are hard to imagine. In this book, we will use analogies to understand worlds with more dimensions than ours. I'd like to start by presenting some analogies that the sphere could use to help the Flatlander understand Spaceland. Later we'll revisit these analogies when we discuss four dimensions.

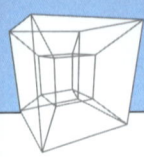

The Fourth Dimension and Beyond

Analogy 1: Right next to you

When the square visited Lineland, it moved next to the line, but the Linelander was completely unaware of the square. There is no way for the Linelander to see the square because it can't look to the side, out of Lineland.

When the sphere visited Flatland, it moved above the square, but the Flatlander was completely unaware of the sphere. There is no way for the Flatlander to see the sphere because it can't look up, out of Flatland.

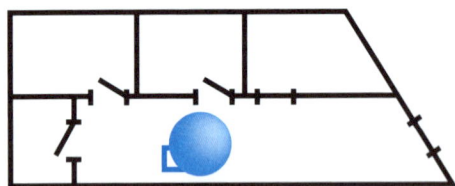

The line thought that the square's voice was coming from inside itself. The square thought that the sphere's voice was coming from inside itself.

Why Can't a Flatlander Imagine Spaceland?

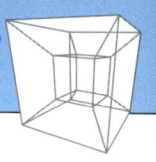

Analogy 2: Seeing entire shapes

The square can see the entire Linelander, but the Linelander sees its neighbors only as points. The Linelander doesn't understand how the square can see both ends of a line at once.

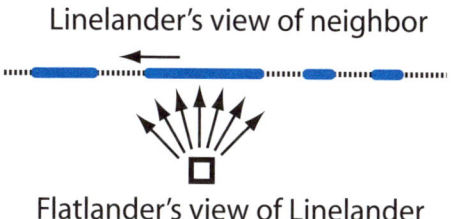

The sphere can see the entire square, but the square sees other Flatlanders only as lines. The square doesn't understand how the sphere can see all four sides of a square at once.

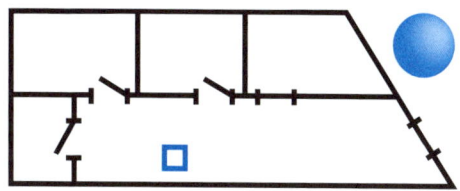

(The sphere has the same view of Flatland as you and I, because it is above Flatland in the 3rd dimension

33

The Fourth Dimension and Beyond

looking down at it, just like you are above the paper looking down at it.)

The sphere can, of course, also see the entire Linelander.

Analogy 3: Seeing inside

Each being can see the insides of objects with fewer dimensions. The square can see the side of the line (which the line thinks of as its insides) and the sphere can see inside the square as well as inside the line.

Lineland is 1-dimensional and is open to the 2nd dimension (and also to higher dimensions). There is nothing the Linelander can do to keep the square from looking at its side.

This seeing inside is not like looking through a window or hole. It is actually looking inside from a different dimension. In Flatland, if the doors and windows are shut, no Flatlander can see inside the house. They can

Why Can't a Flatlander Imagine Spaceland?

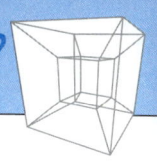

move around the house by moving forward/backward and left/right, but it is impossible for them to rise up, out of Flatland, to look down at the house.

The sphere and we human beings are from Spaceland and we can see all of Flatland and see inside of every building and even inside the Flatlanders. We can see inside Flatland because their world is open to the 3rd dimension (and also to higher dimensions). Since Flatlanders are 2-dimensional, there is no way they can close their world to us.

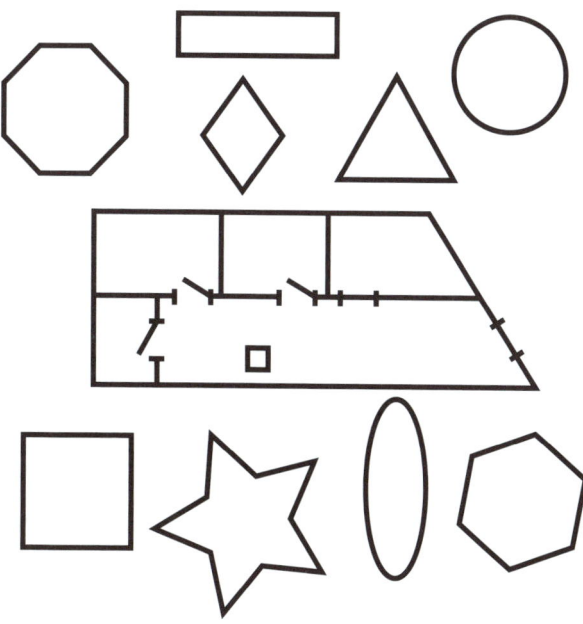

35

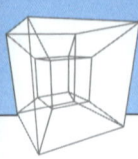

The Fourth Dimension and Beyond

Analogy 4: Making things disappear

The square can move into and out of Lineland. The Linelander sees it suddenly appear and then disappear.

The sphere can move objects into and out of Flatland. To the Flatlander, they suddenly show up and then are gone. We could reach into Flatland from the 3rd dimension and pick up something from within the closed house. The Flatlander would see the object suddenly disappear.

Analogy 5: Cross sections

When an object enters a world that has fewer dimensions, only a cross section of the object exists in that world.

When the square entered Lineland it appeared to the Linelander to be the end of a line, and the Linelander could see only a point.

When the sphere entered Flatland, it showed up as a circle, a circle that started out small and then grew larger. To the Flatlanders a circle looks like a line. They

Why Can't a Flatlander Imagine Spaceland?

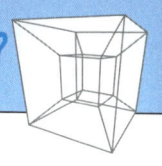

can determine that it is a circle and how large the circle is by moving around and inspecting it.

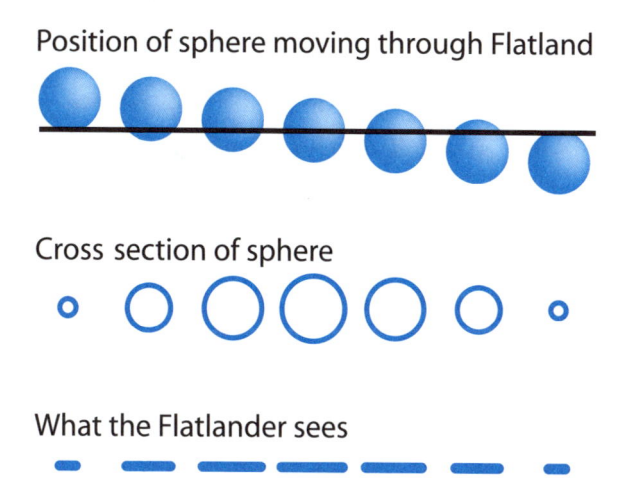

Position of sphere moving through Flatland

Cross section of sphere

What the Flatlander sees

As we have seen, it is easy for the Flatlander to look at and understand Lineland, but it is very difficult for the Flatlander to understand Spaceland. By using these analogies, however, the Flatlander may begin to understand Spaceland even though it can't see Spaceland.

The Flatlander is able to look at the side of the Linelander, something that seems to the Linelander to be impossible no matter how hard the Flatlander tries to explain it. By extending the concept, the Flatlander can begin to understand how a Spacelander can see inside Flatland.

37

The Fourth Dimension and Beyond

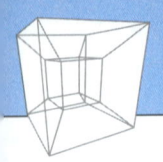

Pretend that you are a Flatlander, living on a sheet of paper. You cannot move, or even look, above or below the piece of paper. Pretend you don't know anything about 3-dimensional Spaceland and try to understand how hard it would be for you to imagine Spaceland. Later in this book, you'll try to imagine four dimensions.

4

DIMENSIONS AND SHAPES

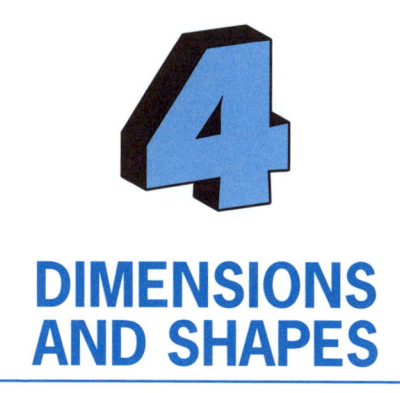

This chapter presents a comparison of the shapes that can be created in the different worlds we have visited. Remember as you read about these shapes that each dimension includes not only the shapes with that many dimensions but also all of the shapes with fewer dimensions. As the number of dimensions increases, both the complexity and the variety of the shapes increase.

Start with a **point**:

0 Dimensions   •

For this demonstration, the point shown here looks like a small circle. In actuality, a point is extremely small. In fact, it is so small that it has no size. Think of it as a

The Fourth Dimension and Beyond

location rather than an object.

Now drag the point to the right:

to make a **line segment**:

1 Dimension

A **line** is 1-dimensional and goes forever in two directions. A **line segment** has two ends and the distance between those ends can be measured. Therefore, a line segment has length. (Line segments are often simply called lines; it is usually obvious whether the word "line" refers to an infinitely long line or a line segment.) A line has length, but, just as a point has no size, a line has no width.

Because dimensions are perpendicular, we will next drag the line segment in a direction perpendicular to the line segment itself, creating two perpendicular line segments on the sides:

Dimensions and Shapes

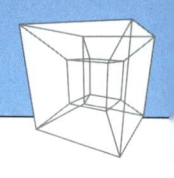

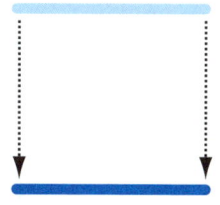

to make a **square**:

2 Dimensions

A flat surface such as a tabletop is a **plane**. Like lines, planes extend forever, and just as a line segment is part of a line, a square is part of a plane. Squares have length and width, but no thickness.

The line segments that make the outside of the square are called the **sides** of the square and all four sides of a square are the same length. The points where the sides come together are called the **corners**. Another word that is used for corner is **vertex**.

The Fourth Dimension and Beyond

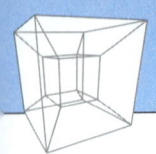

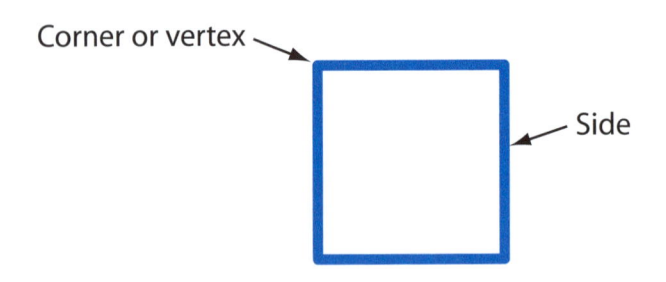

Corner or vertex

Side

The next step is rather difficult because the pages in this book are 2-dimensional. Imagine dragging the square up off the page; drag it straight up, perpendicular to the page. This makes a shape called a **cube**. A cube is a box with six square sides that are all the same size.

Of course, I can't put a cube inside this book. But I can make a drawing of it on this 2-dimensional page by drawing the 3rd dimension at an angle. So the next step is to drag the square down toward the bottom of the page and to the left on the page. Looking at this picture, imagine the square with dashed lines moving up, off the paper, as shown with the arrows:

Dimensions and Shapes

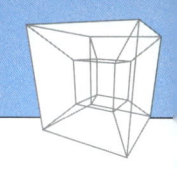

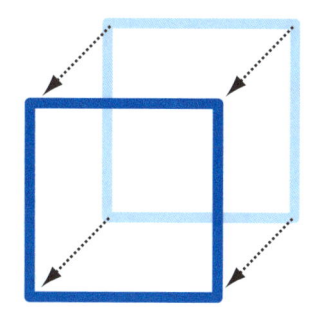

Dragging the square down toward the bottom of the page and to the left on the page creates a 2-dimensional picture of a 3-dimensional cube:

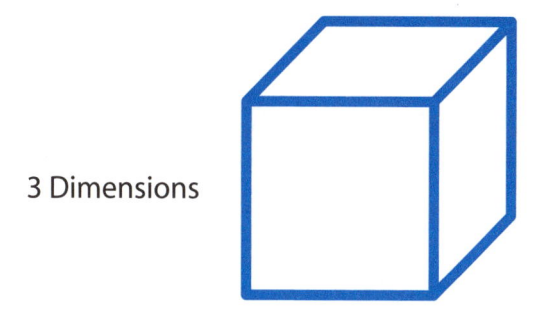

3 Dimensions

A cube has length, width, and height; it encloses a 3-dimensional space.

The squares that the cube is made of are called **faces**. The line segments where the sides of two faces con-

43

The Fourth Dimension and Beyond

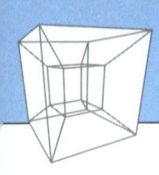

nect are called the **edges** of the cube. The points where three edges meet are the **vertices**. (The plural of vertex is vertices.)

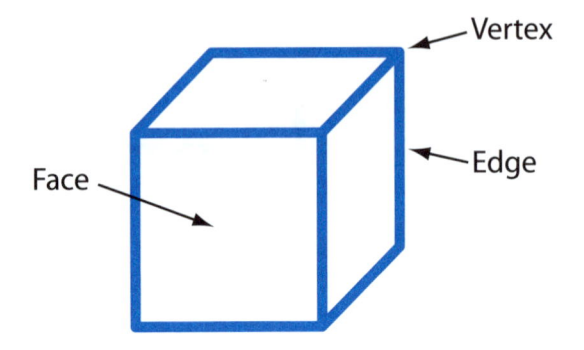

Vertex

Edge

Face

You've probably noticed that two of the faces in this picture are not really squares. They each have four sides but the sides of the squares are not perpendicular at the vertices. This is a result of drawing the 3rd dimension at an angle on the 2-dimensional paper.

In the picture of the cube, you can see only three of the six squares. The other three are on the back of the cube. If you look at an actual box, you will find that you can never see more than three sides at once. You can turn the box around and see the other sides, but you can never see more than three sides at the same time.

Dimensions and Shapes

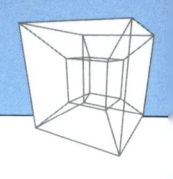

It is possible in the drawing to show the edges and faces that are hidden from view. They are often shown using a different type of line to indicate that they are **hidden lines**. In this picture the hidden lines are shown using dots.

In this picture, try to see the dotted hidden lines as being *behind* the solid lines.

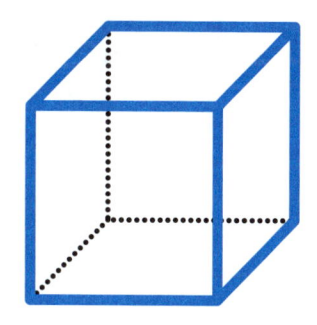

The Fourth Dimension and Beyond

Of course, if a cube is made out of six clear squares, you can look through the faces and see the edges and corners on the other side of the cube. It would look like this:

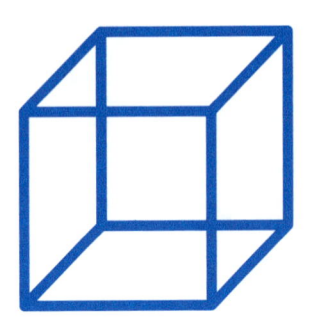

Squares and cubes are convenient shapes to study because it is easy to see the perpendicular dimensions. Remember that there are many shapes in addition to the square that are 2-dimensional (e.g., triangles and ovals). And there are many shapes in addition to the cube that are 3-dimensional (e.g., pyramids, cylinders, and spheres).

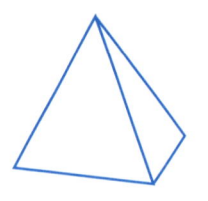

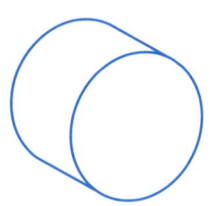

A 4-DIMENSIONAL CUBE

So far, we've looked at a 0-dimensional point, a 1-dimensional line segment, a 2-dimensional square, and a 3-dimensional cube. We made a square by moving a line segment in a perpendicular direction and a cube by moving the square in a perpendicular direction.

Now let's try something that is much harder. Imagine moving the cube in a direction perpendicular to the cube. What does perpendicular to the cube mean? The cube has three dimensions that are perpendicular to one another. We need to move it in a direction that is perpendicular to all three of those dimensions.

Is it hard for you to figure out how to do this? I know it's very difficult for me to imagine.

Suppose there was a 4-dimensional being that lived in a place called **4D-land,** and suppose that this 4D-land-

The Fourth Dimension and Beyond

er offered to come to Spaceland and help us move our cube.

Do you remember how difficult it was for the Flatlander to imagine moving in a direction perpendicular to its world, into the 3rd dimension? But it was very easy for us as humans to see—you just move "up, out of Flatland." It is very difficult for the Flatlander to imagine what a cube looks like, but it is easy for us because we live in 3-dimensional Spaceland.

In a similar way, it would be easy for a 4D-lander to move the cube "up, out of Spaceland." But just as the Flatlander couldn't imagine moving something "up, out of its 2-dimensional Flatland," we can't imagine moving something "up, out of our 3-dimensional Spaceland." And when I say "up, out of Spaceland," I don't mean up above your head. Instead I'm referring to a direction that we can't point to, a direction that's hard to even imagine.

By using this analogy, we can see that, just like it's possible to turn a 2-dimensional square into a 3-dimensional cube, it should be possible to turn a 3-dimensional cube into a 4-dimensional cube. Even though we can't actually make or see a 4-dimensional cube, we

A 4-Dimensional Cube

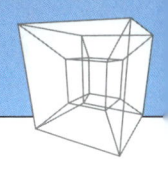

can understand how it would be possible.

Since we are now discussing 4D-land, we need a new pair of directions to use with the new dimension. Let's call these directions **4D-up** and **4D-down** and call the new, 4th dimension **4D-up/4D-down**. In Spaceland, the dimensions forward/backward, left/right, and up/down are all perpendicular to each other. In 4D-land, the dimensions forward/backward, left/right, up/down, and 4D-up/4D-down are also all perpendicular to each other.

So, to make a 4-dimensional cube, all you need to do is drag a 3-dimensional cube 4D-up. But of course we can't do that. So let's draw it instead.

We were able to draw a 3-dimensional cube on our 2-dimensional pages by dragging a square at an angle instead of in a perpendicular direction. Now let's use that same technique to draw a 4-dimensional cube on paper.

The Fourth Dimension and Beyond

Start with a 2-dimensional drawing of a 3-dimensional cube:

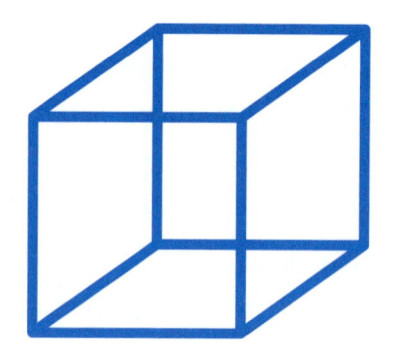

Now we need to show it being dragged 4D-up. To do this, we'll use a different angle than we did when we made the 3-dimensional cube:

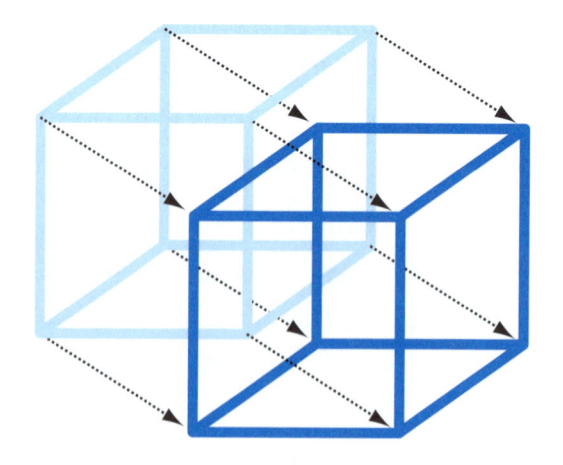

A 4-Dimensional Cube

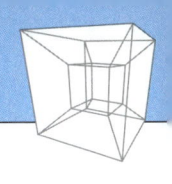

Here is a 2-dimensional drawing of a 4-dimensional cube, made by dragging a 2-dimensional drawing of a 3-dimensional cube. The original 3-dimensional cube is drawn with a solid line. The dashed line shows the position of the second 3-dimensional cube. The dotted lines connect the vertices of the two 3-dimensional cubes to complete the 4-dimensional cube.

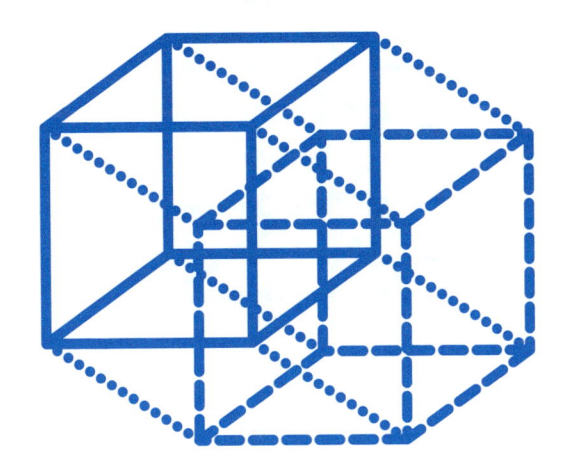

The Fourth Dimension and Beyond

Here is the final 2-dimensional drawing of the 4-dimensional cube. Can you see the original 3-dimensional cube? Can you also see the second 3-dimensional cube?

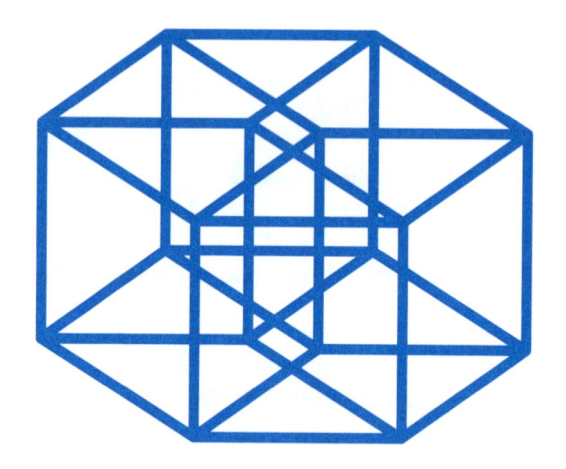

Each of the shapes we have drawn is actually a container and each of these containers encloses a space. The enclosed space has the same number of dimensions as the shape itself.

The 3-dimensional cube is obvious to us—it is a box. We use boxes as containers to hold things. The space inside the boxes we use is obviously 3-dimensional. In Flatland, a square is a container, the equivalent of a cube in our world, and the space inside a square

52

A 4-Dimensional Cube

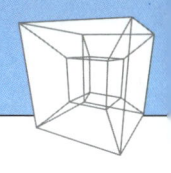

is 2-dimensional. Flatlanders can store 2-dimensional things in a square just as we store 3-dimensional things in cubes. Finally, a line segment in Lineland is a container that can hold lines. The space in a line segment is 1-dimensional.

Each of these containers has a boundary that defines the container and holds items within the container. For example, the top, bottom, and walls of a box (or the square faces of a cube) create the space inside the box and ensure that the things in the box stay in the box.

Have you noticed that each of the shapes is used to form the boundaries for the shape of the next higher dimension?

Points form the boundaries (the ends) of a line segment:

The Fourth Dimension and Beyond

Lines form the boundaries (the sides) of a square.

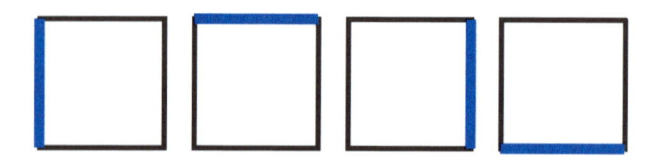

Squares form the boundaries (the faces) of a cube:

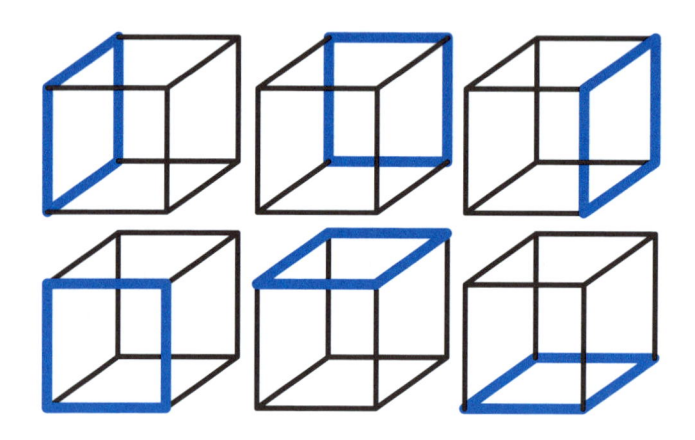

To summarize:

    A line is bounded by two points.

    A square is bounded by four lines.

    A cube is bounded by six squares.

Do you see the progression? Do you know now what

A 4-Dimensional Cube

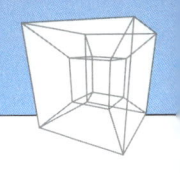

the outside of a 4-dimensional cube is made of?

A 4-dimensional cube is bounded by eight 3-dimensional cubes:

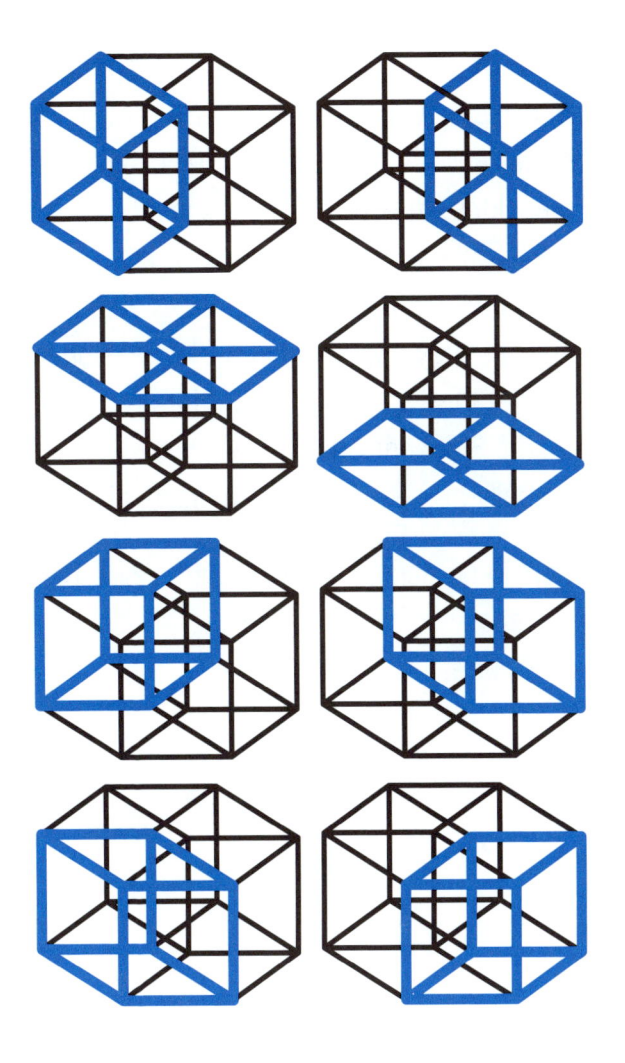

55

The Fourth Dimension and Beyond

In 4-dimensional space, eight 3-dimensional cubes are the faces of the 4-dimensional cube.

In these drawings, the 3-dimensional cubes overlap one another and appear to be in the same space. Also, some of the squares that make the sides of the cubes are distorted. These problems occur when 4-dimensional and 3-dimensional objects are drawn in two dimensions. In a real 4-dimensional cube, the 3-dimensional cubes would not overlap, just like the six squares in a 3-dimensional cube don't really overlap each other as they appear to in a 2-dimensional drawing. Although they are distorted in the drawings, each of the eight 3-dimensional cubes has six faces that would actually be squares in 4-dimensional space.

A 4-dimensional being can, of course, see the eight cubes as easily as we can see the six squares that make up a cube.

A 4-Dimensional Cube

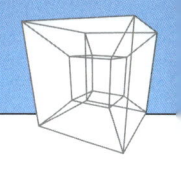

This table shows what we have learned so far:

Dimensions	Shape	Number of boundary pieces	Shape of the boundary pieces
1	Line	2	Points
2	Square	4	Lines
3	Cube	6	Squares
4	4-D cube	8	Cubes

And we can figure out that

a 5-dimensional cube is made of 10 4-dimensional cubes.

a 6-dimensional cube is made of 12 5-dimensional cubes.

a 158-dimensional cube is made of 316 157-dimensional cubes.

Where did that last one come from? Look back at the numbers in the table; you will see that to figure out how many parts are needed to make a shape, you just multiply the number of dimensions by two: 158 × 2 = 316. And to figure out what those parts are, subtract one from the number of dimensions: 158 − 1 = 157.

The 4-dimensional cube has been given a special

57

The Fourth Dimension and Beyond

name: it is called a **tesseract**.

Another name that is used is **hypercube**. The prefix *hyper* means above or over. In this case, it means "more dimensions than we are used to": in other words, more than three dimensions. As you read more about hyperdimensions, you will see *hyper* used a lot.

Since *hypercube* actually means a cube with four, five, six, one thousand, or any other number greater than three dimensions, a more accurate name for the cube we've been discussing is **4-dimensional hypercube** or **4D-hypercube**. And oftentimes that is shortened to **4D-cube** or even **4-cube**.

This may sound a little strange, but we could call the point, line segment, and square a 0D-cube, a 1D-cube, and a 2D-cube:

0D-cube   1D-cube   2D-cube   3D-cube   4D-cube

Later in this book we'll see how this consistent naming is useful.

58

WHAT IS 4D-LAND LIKE?

As we visited Pointland, Lineland, and Flatland, it probably occurred to you that those worlds are pretty boring compared to our Spaceland. In fact, to a Flatlander, Pointland and Lineland would be much too simple compared with its homeland.

The conclusion we can draw from this is that as the number of dimensions increases, things get more and more interesting. Pointland, after all, consists of nothing but one extremely tiny point. Lineland is just a bunch of lines in a row.

Flatland is a big improvement over Pointland and Lineland. Whereas Pointland has exactly one point and Lineland has lines of various lengths, Flatland has squares and triangles and circles and ovals and lots of other shapes in many different sizes.

The Fourth Dimension and Beyond

In Pointland there is nothing for the point to look at, and all a Linelander can see is the ends of its two neighbors, which look like points. Flatlanders can move around in their world and see all of the other shapes, although those shapes look like lines.

Consider how much more we have in Spaceland. In addition to everything that the Pointlander, Linelanders, and Flatlanders have, we have 3-dimensional objects. We can move in more directions than they can. We can see more than they can.

Well, if you think Spaceland is great, just think of what 4D-land must be like. They have everything we have plus everything that another dimension gives them. And, of course, 4D-land would be boring to a 5D-lander, and 5D-land would be boring to a 6D-lander. I wonder what 100 dimensions would be like!

Imagine that a 4D-cube came to you to explain a world with four dimensions. Remember how easy it was for us to understand what the sphere was trying to explain to the Flatlander? Now, we'll understand why it was so hard for the Flatlander to comprehend.

We'll start by looking again at the analogies in chapter 3.

What is 4D-Land Like?

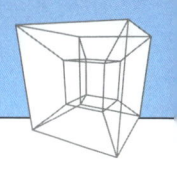

Analogy 1: Right next to you

Just as the square was right next to the line without the line knowing, and the sphere was just above the square without the square knowing, there could be a 4D-being right next to you and you wouldn't know it.

It could be just inches away, right there 4D-up above you—or 4D-down below you—and watching everything you do. As long as it didn't enter your 3D-world, you wouldn't be able to see it or detect its presence because it would be positioned away from you in the 4th dimension.

If it spoke to you, it would seem as if the voice was coming from inside you.

Analogy 2: Seeing entire shapes

Just as the square could see the entire line, and the sphere (and we) could see the entire square, the 4D-being can see all of you at once—your front, back, both sides, the top of your head, and the bottom of your feet all at the same time.

When we look at Flatland, we see all the sides of every

61

building in one view. The 4D-lander can see every side of every building in one of our cities, without having to move. It can see all of the sides of your house at once.

Analogy 3: Seeing inside

Just as we can see inside a square, a 4D-being would be able to see inside our houses and see everything inside every room, inside every closet, and inside every cabinet. It could even see inside a safe.

It would not see inside by looking through a window or a hole or an open door. It would see everything because our world is completely open to the 4th dimension in a way that is similar to Flatland being completely open to us. Just as we can see inside a Flatlander's house from the 3rd dimension, a 4D-lander could see inside our house from the 4th dimension. Just as it is impossible for a Flatlander to close up its house to us because it can't move in the 3rd dimension, it is impossible for us to close up our house to the 4D-lander because we can't move in the 4th dimension.

The 4D-being could also see inside us. It could see all

What is 4D-Land Like?

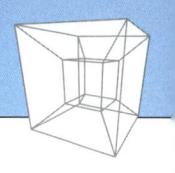

of our organs and could see what we ate for lunch. A 4D-surgeon could operate on us without cutting our skin.

Analogy 4: Making things disappear

Just as the sphere could move objects into and out of Flatland, a 4D-lander could move items into and out of our world. It could reach into your kitchen, pick up the sandwich off your plate, and lift it 4D-up, out of your world. You wouldn't see the sandwich move through the window or out the door. You would just see it disappear.

We've learned that we could lift something that is in a Flatlander's closed house up into the 3rd dimension and set it down outside the house.

A 4D-lander could lift something from a sealed box in a closet in your locked house up into the 4th dimension and set it down in your yard without going through a door or window or making a hole in the roof.

The Fourth Dimension and Beyond

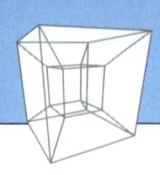

Analogy 5: Cross sections

When the sphere moved through Flatland, the Flatlanders saw a circle that was first small, then grew larger, and then shrank to a small size again and disappeared.

A **cross section** is the shape that is formed by cutting an object. The cross section of a sphere is a circle. (The cross section has one less dimension than the original object.)

If a 4D-sphere were to move through your living room, you'd suddenly see a small 3D-sphere—a ball—appearing to float in the air, and then you'd see it grow larger and larger. After reaching its largest size, it would start to shrink, eventually becoming tiny and disappearing. The cross section of a 4D-sphere is a 3D-sphere, just like the cross section of a 3D-sphere is a circle.

Are you ready for some new analogies now?

Analogy 6: Rotation

Is it possible for a person with a mole on the left cheek to be rotated or "turned around" so that the mole is on the right side?

What is 4D-Land Like?

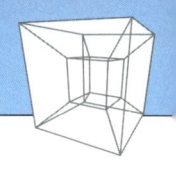

Let's start with a Flatland analogy. Here's a picture of a happy square:

Can you figure out how to turn it around so it is facing the other direction but still has the top of its head toward the top of the page? Spinning it around doesn't work:

But if we pick it up off the paper and flip it over, it will face the other direction:

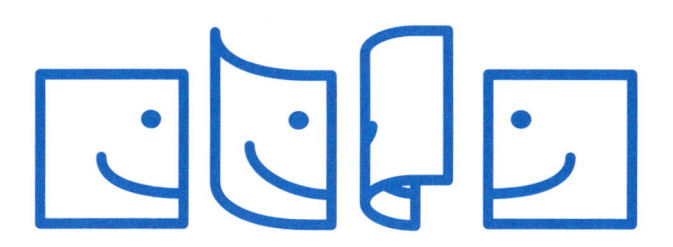

65

The Fourth Dimension and Beyond

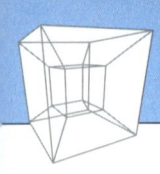

The act of "flipping it over," though, is actually rotating it through the 3rd dimension. We had to pick it up off the paper, rotate it, and then put it back down on the paper. There is no way a Flatlander could do this without help from a Spacelander.

Now do you see how to get the mole to the other side of the face? We just need a 4D-lander to pick the person up and rotate him or her through the 4th dimension. If this were to actually happen, the person would be completely reversed and all of the organs inside would be switched left-to-right.

Analogy 7: Making spheres

To make a sphere, start by selecting a single point. This point will be the center of the sphere. Imagine connecting a string to this center point and stretching it out tight. Now move the string in every direction possible, always keeping the string pulled tight. The shape that the end of the string makes is a sphere. Every point on a sphere is the same distance from the center point; this distance is called the **radius** of the sphere.

Reread the previous paragraph and notice that it didn't

What is 4D-Land Like?

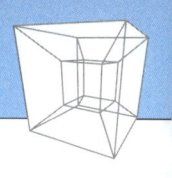

specify how many dimensions there are. At the end of chapter 5, we defined cubes in many different dimensions, using names like "2D-cube." Let's do the same thing with a sphere.

In one dimension, we have a line. Pick a point (**C**) on the line for the center of our circle:

C

Now tie the imaginary string to the center point and stretch it tight. In one dimension, we can only move it to the left and right of the center point:

1D-sphere

C

It's not much, is it? But it is a 1-dimensional sphere according to our definition of a sphere.

For two dimensions, the string can move forward and backward as well as left and right. In addition, it can move to many different combinations of those directions as long as it is pulled tight.

The Fourth Dimension and Beyond

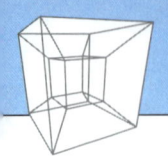

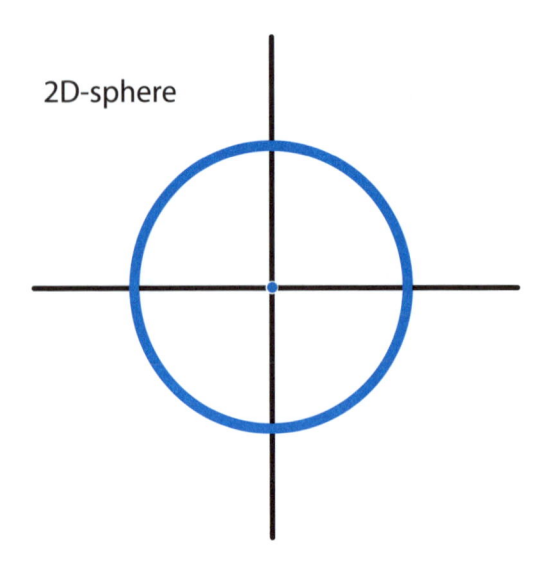

2D-sphere

We usually call this a **circle**.

In three dimensions, of course, a sphere is made by pulling the string in all possible combinations of the forward, backward, left, right, up, and down directions.

What is 4D-Land Like?

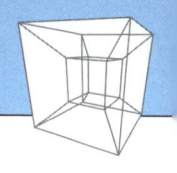

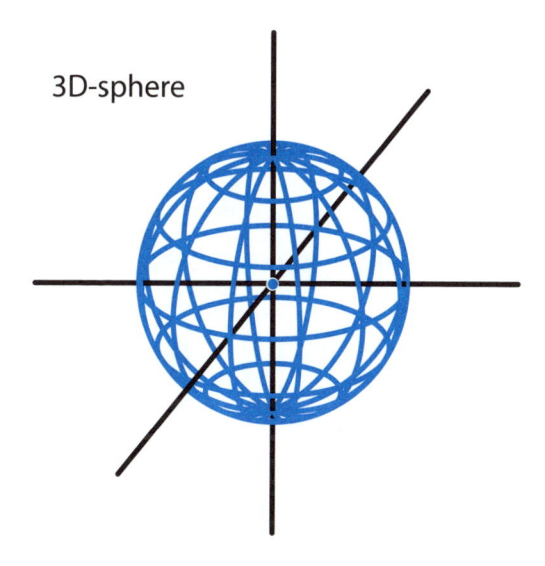

3D-sphere

I can't draw a 4D-sphere and none of us Spacelanders are able to see a real one, but we know how to make one—just move the string through four different dimensions to trace out the shape. And, although we can't see it, we know that it will be curved.

Here's another idea. You can make a 3D-shape that is similar to a sphere by cutting out circles from cardboard and stacking them up. The bottom piece should be a small circle, the next one slightly larger, and so on, until you have a circle the size of the 3D-sphere that you want to make. Then the circles start getting

69

The Fourth Dimension and Beyond

smaller again until you have a small one at the top.

Stack them up carefully and glue them together and you'll have something that resembles a 3D-sphere.

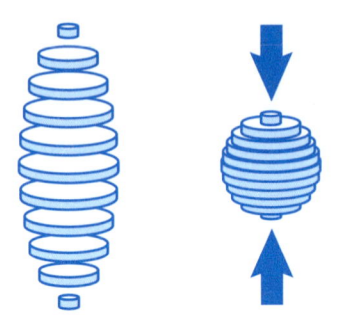

So how can you build a 4D-sphere? Make a series of 3D-spheres from small to large and back to small again.

And then stack them up in the 4th dimension. Just as 2-dimensional circles stacked up make a sphere, stacking up the 3D-spheres in the 4th dimension would make something close to a 4D-sphere.

HOW MANY PARTS DOES A CUBE HAVE?

We have seen how simple shapes are used to make more-complicated shapes and how those shapes are used to make even more-complex shapes. Let's look a little more closely at these shapes to see how they help us understand higher dimensions.

Remember from chapter 4 that a vertex is a point in a shape at which ends of the line segments connect to each other. For example, in a square, the ends of 2 line segments join to make each corner of the square. These corners are the vertices of the square; a square has a total of 4 vertices.

A cube also has vertices. In a cube, three line segments come together at each vertex, and there are a total of 8 vertices in a cube. (Remember that when we made a cube by dragging a square, we had 2 squares with 4

The Fourth Dimension and Beyond

corners each, a total of 8 corners. When we connected the corresponding corners of the 2 squares, we added a new line segment to each corner, so the number of line segments meeting in each corner is 3.)

So far we have this:

	Square	Cube
Dimensions	2	3
Vertices	4	8
Line segments at vertex	2	3

Now think about a simple line segment. It doesn't have any corners, but it does have two ends. These ends can be thought of as vertices that each have one line segment:

	Line	Square	Cube
Dimensions	1	2	3
Vertices	2	4	8
Lines at vertex	1	2	3

It's even stranger to think about a point having corners, but look at the numbers we have so far in the table:

A cube (3 dimensions) has 8 vertices.

A square (2 dimensions) has half as many as a cube: 4.

How Many Parts Does a Cube Have?

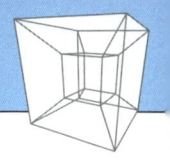

A line segment (1 dimension) has half as many as a square: 2.

So, a point (0 dimensions) should have half as many as a line segment: 1.

And, since there are no line segments in a point, the number of line segments at the point's vertex is 0. (Think about the point as a vertex where no line segments meet.)

	Point	Line	Square	Cube
Dimensions	0	1	2	3
Vertices	1	2	4	8
Lines at vertex	0	1	2	3

Next, let's go beyond our 3D-world. Look at the table we've made. Can you determine how many vertices a 4D-cube has? Each time the number of dimensions increases by one, there are twice as many vertices. Following the numbers in the chart, we can see that a 4D-cube has 16 vertices.

And how many line segments meet at each vertex? This one is easy: 4.

	Point	Line	Square	Cube	4D-cube
Dimensions	0	1	2	3	4
Vertices	1	2	4	8	16
Lines at vertex	0	1	2	3	4

The Fourth Dimension and Beyond

Now, without looking ahead, figure out the numbers for a 5D-cube.

Ready? Do you have the answers?

	Point	Line	Square	Cube	4D-cube	5D-cube
Dimensions	0	1	2	3	4	5
Vertices	1	2	4	8	16	32
Lines at vertex	0	1	2	3	4	5

Using the naming that we learned about at the end of chapter 5, the table looks like this:

	0D-cube	1D-cube	2D-cube	3D-cube	4D-cube	5D-cube
Dimensions	0	1	2	3	4	5
Vertices	1	2	4	8	16	32
Lines at vertex	0	1	2	3	4	5

We could work forever making this chart larger and larger. Is there a way to tell someone how to figure out the answers for any number of dimensions without having to make a *really* huge chart for them?

The number of line segments is obvious; it's the same as the number of dimensions.

To figure out the number of vertices, we just kept multiplying by 2. In fact, if you multiply 2 × 2 × 2 . . . and write down as many 2s as there are dimensions,

How Many Parts Does a Cube Have?

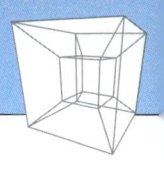

that will tell you how many vertices there are. In 2 dimensions, for example, 2 × 2 is 4. A square has 4 vertices. In 5 dimensions, 2 × 2 × 2 × 2 × 2 is 32. A 5D-cube has 32 vertices.

There is a shortcut way to write this. 2^5 means 2 × 2 × 2 × 2 × 2. The 2 is called the **base**; the ^{5} is the **power** or **exponent.** 2^5 is read, "2 raised to the 5th power," often shortened to "2 to the 5th." Notice that the exponent is written smaller than the base and it is positioned higher than the base.

Now there's one more trick. We can use a letter to represent the number of dimensions. I'm going to use the letter "n" but any letter could be used. The letter "n" will represent any of the counting numbers (0, 1, 2, 3, and so on). Notice carefully how "n" is used in the next sentence:

> An **n**D-cube has **n** dimensions, 2^n vertices, and **n** line segments at each vertex.

You can pick a number and replace "n" with your number. For example, if you pick 4, the sentence changes like this:

> A **4**D-cube has **4** dimensions, 2^4 vertices, and **4**

The Fourth Dimension and Beyond

line segments at each vertex.

2^4 is the same as $2 \times 2 \times 2 \times 2$, which is equal to 16. So the sentence can be rewritten:

A **4**D-cube has **4** dimensions, **16** vertices, and **4** line segments at each vertex.

We'll now add one more shape to our table, the nD-cube. (By the way, 2^0 is equal to 1 and 2^1 is equal to 2.)

	0D-cube	1D-cube	2D-cube	3D-cube	4D-cube	5D-cube	nD-cube
Dimensions	0	1	2	3	4	5	n
Vertices	1	2	4	8	16	32	2^n
Lines at vertex	0	1	2	3	4	5	n

You'll notice that we could leave out dimensions 0 through 5 because the nD-cube works for all of them. It's often easier to understand, though, if a table is made showing the first several entries and then the one that shows how to calculate the answer for any number.

Now you see why we called a point a 0D-cube, a line segment a 1D-cube, and a square a 2D-cube.

Using the entries in the table for the nD-cube, can you answer the following questions?

How Many Parts Does a Cube Have?

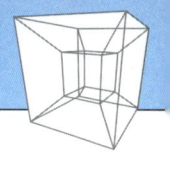

Question: How many vertices are in a 12D-cube?

Answer: 2^{12}, which is 4096.

Question: How many line segments meet at each vertex?

Answer: 12.

Do you understand how mathematical notation can be used to easily describe cubes in any number of dimensions?

If you are patient and careful, you can make a chart that lists, for each shape, how many of each simpler shape it contains:

This shape: Has:	0D-cube	1D-cube	2D-cube	3D-cube	4D-cube
0D-cubes (vertices)	1	2	4	8	16
1D-cubes (lines)	0	1	4	12	32
2D-cubes (squares)	0	0	1	6	24
3D-cubes (cubes)	0	0	0	1	8
4D-cubes (4D-cubes)	0	0	0	0	1

If you look under the shape "3D-cube," for example, you will see that it has 8 vertices, 12 lines, 6 squares, and, of course, 1 cube.

The Fourth Dimension and Beyond

In chapter 4 we started with a point and then made a line segment, a square, and a cube; and in chapter 5 we made a 4D-cube, as summarized in this table:

Line segment	made from 2 points (vertices)
Square	made from 4 line segments
Cube	made from 6 squares
4D-cube	made from 8 cubes

You can see those numbers, 2, 4, 6, and 8, highlighted in this table:

This shape: Has:	0D-cube	1D-cube	2D-cube	3D-Cube	4D-cube
0D-cubes (vertices)	1	**2**	4	8	16
1D-cubes (lines)	0	1	**4**	12	32
2D-cubes (squares)	0	0	1	**6**	24
3D-cubes (cubes)	0	0	0	1	**8**
4D-cubes (4D-cubes)	0	0	0	0	1

This idea can be rewritten, again using **n** to represent any counting number, in precise mathematical terms as follows:

An **n**D-cube is made from (2 x **n**) (**n** – 1)D-cubes.

For example, if you want to make a 4D-cube you would write:

How Many Parts Does a Cube Have?

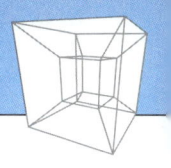

A **4**D-cube is made from (2 x **4**) (**4** − 1)D-cubes,

which becomes this:

A **4**D-cube is made from **8 3**D-cubes.

Look back at the chart and you will see that this is correct.

By using a letter to represent any number and by designing a consistent way to name the objects (0D-cube, 1D-cube, 2D-cube, 3D-cube . . .), mathematicians can express important ideas in short, simple ways.

In previous chapters, we learned about worlds with more than three dimensions by using analogies, and we discovered how difficult it is to understand what a 4-dimensional shape looks like.

In this chapter, we learned how to describe those worlds mathematically. It is difficult for us to imagine worlds with many dimensions, but using mathematics it is easy to describe, in precise detail, ideas that challenge our imagination.

8

HOW CAN WE TRY TO IMAGINE 4D-LAND?

In this chapter, I'll show several ideas that may help you understand four dimensions better. I can't promise that you'll be able to visualize 4-dimensional shapes, but maybe you'll at least see how four dimensions could be possible.

SHOW ME THE NEXT DIMENSION

In Flatland, there are two perpendicular dimensions and we can show them like this:

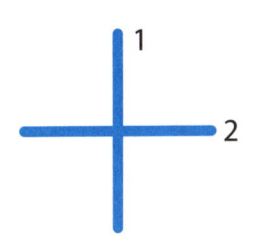

The Fourth Dimension and Beyond

If we asked our Flatland friend, the square, to show us a 3rd dimension that is perpendicular to those dimensions, it would have no idea how to do it. It might even reply, "You're crazy; that is impossible." (Imagine that you are the square and remember that you are stuck on this sheet of paper. You don't know anything about up and down. You can't move up above the paper. You can't even look up above the paper.)

But when you, as a 3-dimensional human, are asked to do it, you would say, "Sure, that's easy. It just comes straight up, out of the paper. In fact it goes straight down through the paper, too."

That was easy, right? Why was it so hard for the square? Because it can't move into or see into the 3rd dimension.

The girl in this picture is holding three sticks that are fastened together and represent our three dimensions.

How Can We Try to Imagine 4D-Land?

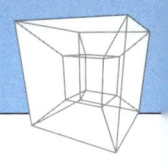

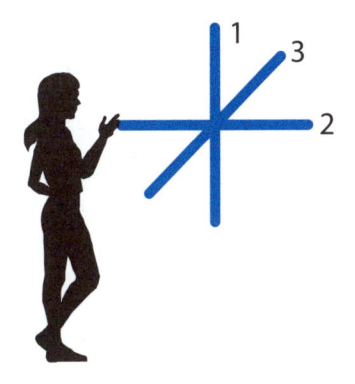

Each stick is perpendicular to each of the other two sticks so they are mutually perpendicular. (They don't appear to be perpendicular on the page because it is a 2-dimensional drawing. Imagine that one end of stick 3 comes straight up out of the paper and the other end goes straight down below the paper.)

Assume that you are holding these sticks and I ask you to show me a 4th dimension that is perpendicular to all three of the sticks. Can you do it? Can you imagine how to do it?

I can't imagine it, but I believe that a 4-dimensional person could do it because I see how we, as 3-dimensional people, can add a 3rd dimension to the two dimensions in Flatland, which is just as hard for a Flat-

The Fourth Dimension and Beyond

lander to imagine.

Wouldn't it be nice to have a 4-dimensional person add the fourth stick so we could see how to do it? But think about this for a minute. Would you be able to see the fourth stick that was added perpendicular to our three dimensions?

Just as the Flatlander can never see the 3rd dimension we added, we could never see the fourth stick that the 4-dimensional person added to our three dimensions. It is not possible for us to see into the 4th dimension.

LOOKING AT CUBES

If you hold up a clear cube (a 3D-cube) and look straight at the middle of one of the faces, what do you see?

How Can We Try to Imagine 4D-Land?

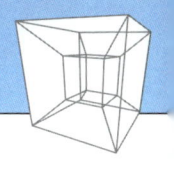

There are two squares. The larger one is the edges of the face of the cube that is closest to you. When you look through that face, you will see the square that is on the other side of the cube. This square appears to be smaller than the first square because it is farther away. Because you are looking straight into the cube, the square that looks smaller appears to be inside the square that looks larger.

A line connects each vertex of the front square to the corresponding vertex of the square in back. This straight-on view of a clear 3D-cube is a 2-dimensional figure that looks something like this:

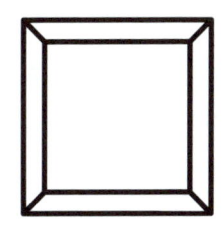

You can see the same effect by looking inside an empty box.

You know that a 3D-cube is made of 6 square faces. I've already mentioned the larger front square and the smaller square in back. The other four are on the two

The Fourth Dimension and Beyond

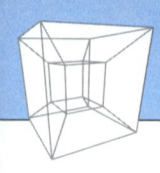

sides, the bottom, and the top. In this picture, the square face on the right side is highlighted:

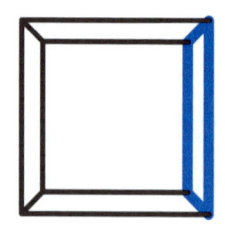

The side does not look like a square in this 2-dimensional image because of the way it is positioned in relation to our eyes. But you can see that it has four sides and four vertices.

The 2-dimensional image of the 3D-cube is a square within a square. Now imagine a 4D-person looking straight at a clear 4D-cube. By analogy, the image that the 4D-person sees would be a 3D-cube within a 3D-cube. (Remember that the 4D-cube is made of eight 3D-cubes.)

Again using analogy, the 3D-person sees a 2-dimensional image of a 3D-cube; the 4D-person would see a 3-dimensional image of the 4D-cube. And, just as edges (lines) connected the vertices of the squares, there would also be edges between the vertices of

How Can We Try to Imagine 4D-Land?

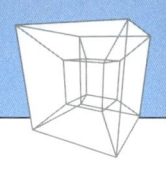

the 3D-cubes in the image of the 4D-cube. This boy is holding a 3-dimensional model of the image that the 4D-person would see:

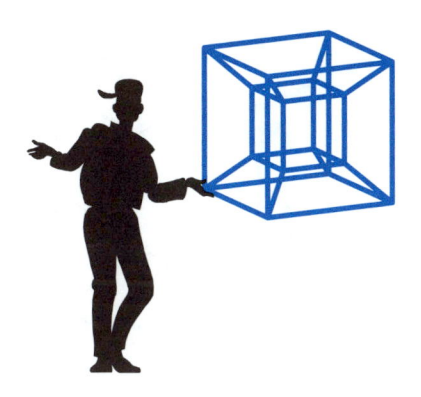

A 4D-cube is made of eight 3D-cube faces. I've already mentioned the larger cube:

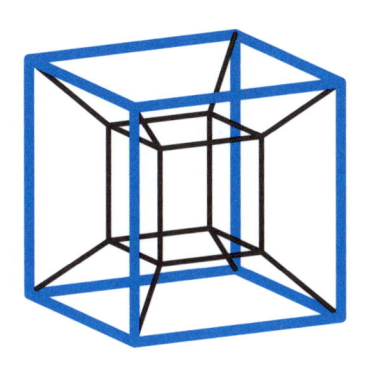

and the smaller cube inside it. The other six are on

87

The Fourth Dimension and Beyond

the two sides, the bottom, the top, the front, and the back. In this picture, the 3D-cube face on the left side is highlighted:

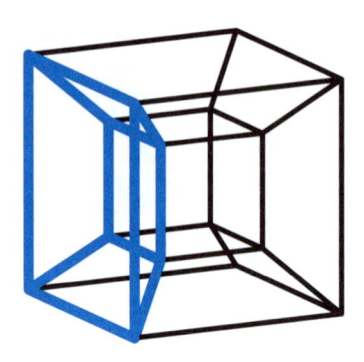

Just as the square faces of the 3D-cube didn't look square, these 3D-cubes also are distorted in the image.

Do you see how the cube-within-a-cube is analogous to the square-within-a-square? Can you imagine what you would see if you were a 4D-person looking straight into a clear 4D-cube? It's hard. Remember that a 4D-person looking at a 3D-cube would see all six sides at once.

How Can We Try to Imagine 4D-Land?

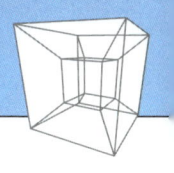

HOW TO FOLD A CUBE

Imagine cutting a 3D-cube open along some of its edges and unfolding it. We may get a shape that looks like this:

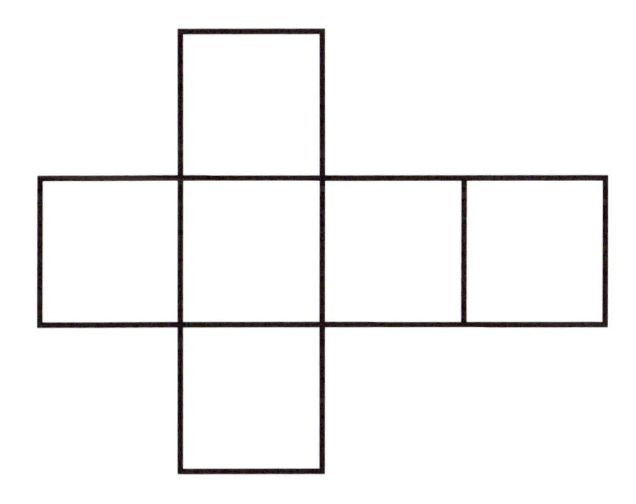

(Depending on which edges we cut, several different shapes are possible.)

An unfolded 3D-cube is a 2-dimensional object. Let's take this unfolded 3D-cube and show it to our friends in Flatland. What will they see? By moving around it and observing its edges and angles, they will be able to conclude that the shape is made of five 2D-cubes (squares). They will probably also realize that there is

89

The Fourth Dimension and Beyond

space for a sixth 2D-cube inside the shape, but they won't be able to tell if there is actually a 2D-cube there or just a 2D-cube-shaped hole.

Now we might try explaining to our 2-dimensional friends that this shape can be folded into a 3-dimensional cube. When we explain the math to them, it seems believable but when they try to actually fold it, they don't know what to do.

The first step is to fold two of the 2D-cubes up so their edges connect:

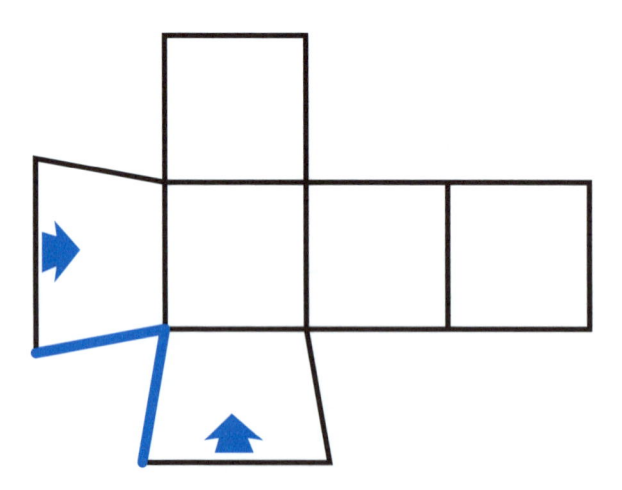

How Can We Try to Imagine 4D-Land?

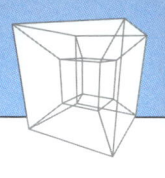

They may try to move one of the squares like this:

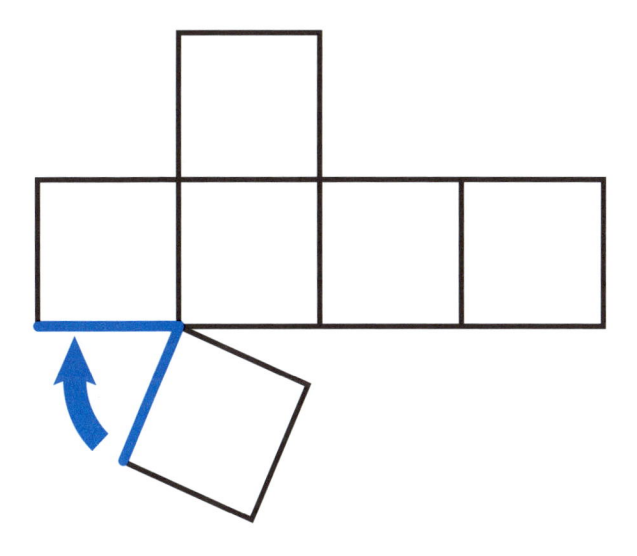

And we would say, "No, don't pull it apart. Bend the two squares *upward* (into the 3rd dimension) until their edges touch."

Of course, they would never succeed because they do not have the ability to move objects upward, into the 3rd dimension.

Suppose, now, that a 4D-lander cut open and unfolded a 4D-cube and put it into our 3-dimensional world. (A 4D-cube unfolds into a 3-dimensional object.) Here is one of the many possible shapes that we could see,

The Fourth Dimension and Beyond

depending on how it was cut and unfolded:

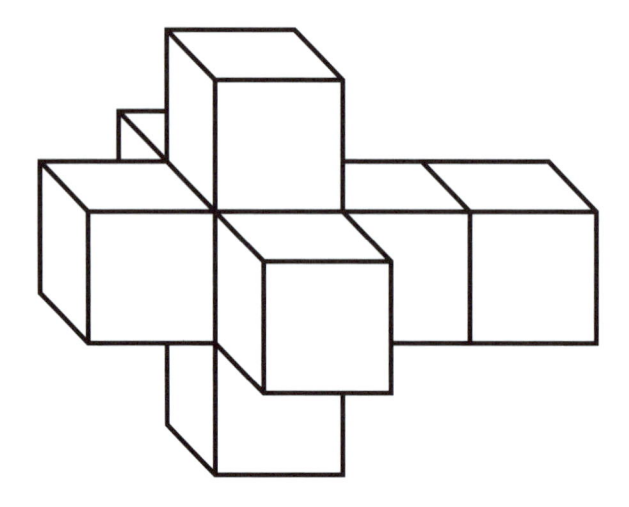

We could inspect the object and determine that it was made of seven 3D-cubes and that there is a place in the middle where there could either be an eighth 3D-cube or an empty 3D-cube-shaped space. (Our 4-dimensional friend could, of course, see the eighth 3D-cube in the middle just as we can see the 2D-cube that is in the middle of the unfolded 3D-cube.)

Our friend explains that this object can be folded into a 4D-cube. While a 3D-cube is made of six 2D-cubes connected at their *edges*, a 4D-cube is made of eight 3D-cubes connected at their *faces*. The first step is to

How Can We Try to Imagine 4D-Land?

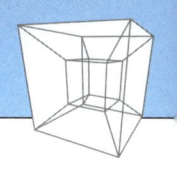

fold it so that two of the faces come together.

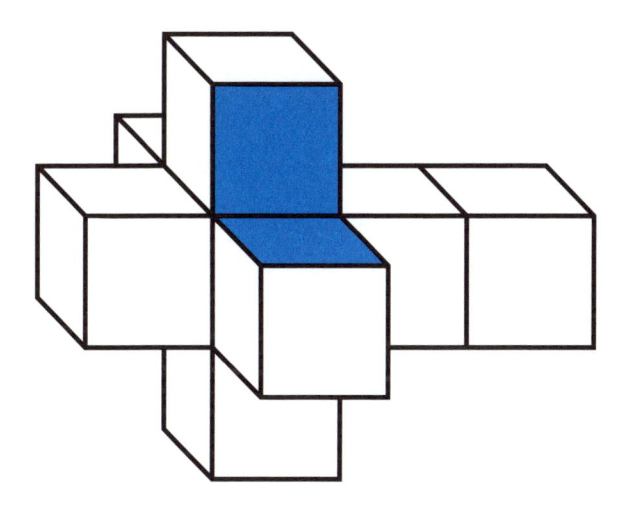

We might try moving one of the cubes like this:

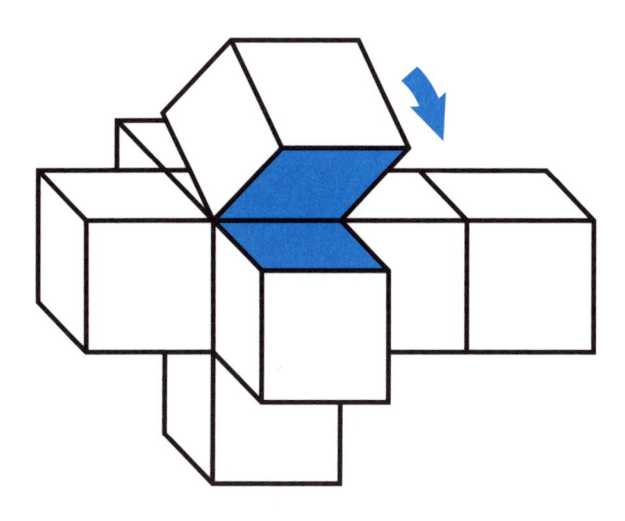

93

The Fourth Dimension and Beyond

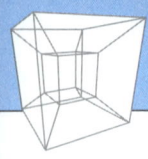

But our 4-dimensional helper would say, "No, don't pull it apart. Bend the 3D-cubes *4D-upward* until their faces touch." We need to connect the faces without separating or bending any of the 3D-cubes.

Of course, we will never succeed because we are unable to move objects into the 4th dimension.

Being curious, we ask our 4-dimensional teacher to do it for us. What will we see when the unfolded 4D-cube is folded back together? To answer that, let's go back to Flatland and the unfolded 3D-cube. If a Flatlander watches us fold the 3D-cube, what will it see?

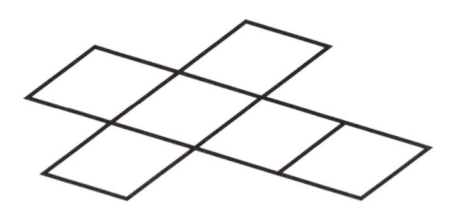

As we lift one of the 2D-cubes up into the 3rd dimension, the Flatlander will see it disappear.

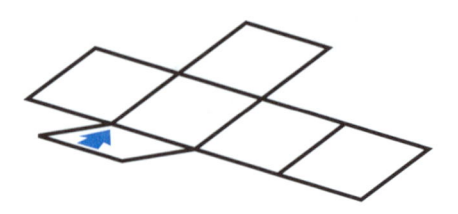

How Can We Try to Imagine 4D-Land?

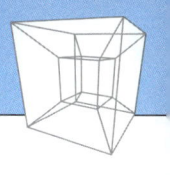

As we continue folding the cube, each of the 2D-cubes will disappear as we fold it up, except for the one in the middle. When we finish folding the cube, all that will remain in Flatland will be a single 2D-cube, a square.

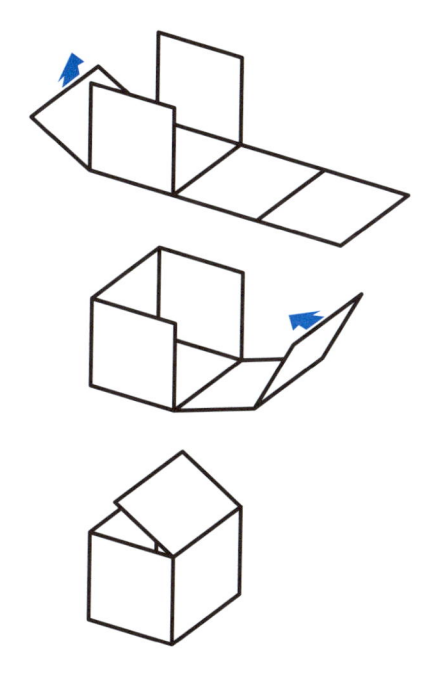

The analogy for the 4D-cube is obvious, right? As the 4D-lander folds the 3D-cubes together, they will move into the 4th dimension and disappear. When the folding is complete, the only thing left will be a single 3D-cube.

The Fourth Dimension and Beyond

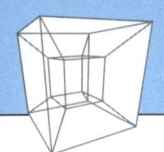

That is rather disappointing, isn't it? But at least we were able to see an unfolded 4D-cube. And even though we can't see an actual 4D-cube, we've learned a little more about them.

ISN'T THE 4TH DIMENSION TIME?

When Albert Einstein developed the theory of relativity, he realized that time could be treated like another dimension. The dimensions we have been learning about are called **spatial** dimensions because they describe locations in the space of our world. The dimension of time can be treated like the spatial dimensions from a mathematical point of view. It's easier to use the same calculations on space and time than to treat them separately. This combined, or unified, system of spatial and time dimensions is called **spacetime**.

In reality, time is different because we can only move forward in time, and we can't stop moving in time, unlike the spatial dimensions, where we can move in either direction on each dimension or stay in one location. Also, spatial dimensions are measured as

The Fourth Dimension and Beyond

distances using measurements like meters, feet, and inches, whereas time is measured using years, hours, seconds, etc. In spite of these differences between space and time dimensions, treating them as dimensions of one space simplifies the theory of relativity.

Einstein was working with the three spatial dimensions of Spaceland, so he added time as the 4th dimension. In this book, we have been exploring worlds with fewer than three dimensions and with more than three dimensions. Time can be added to any of those worlds as another dimension in the same way Einstein added it to three dimensions.

In Flatland, for example, time would be the 3rd dimension and in 5D-land, time would be the 6th dimension. In other words, there is a spacetime for Flatland and a spacetime for 5D-land.

10

IS THIS USEFUL?

"Is this useful?" One answer to that question is another question, "Does it matter?" Sometimes things are worth figuring out, learning about, and understanding just because it's interesting and challenging. Anything that expands our minds and causes us to think in new ways will improve our thinking and problem-solving abilities. And that can be useful in many ways throughout our lives.

Another answer is, "Maybe." We never know when someone will use new ideas in ways that no one anticipated. There are many examples throughout history where discoveries made by one person are later used by someone else to solve a problem or create something new.

And the actual answer is, "Yes. In fact, the ideas and techniques developed by studying multiple dimensions are used in many different ways." In this chapter, we'll take a brief look at a few examples.

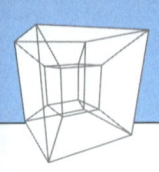

HYPERCUBE COMPUTER

One of the techniques that computer scientists have used to improve the capabilities of computers and to reduce their costs is to connect many (hundreds or thousands) of small, low-cost computers together. Programs are developed that divide the work up into parts. Those parts are sent to the individual small computers, and all the computers work on their parts of the problem at the same time.

This technique is called parallel processing, because the small computers are working at the same time (or "in parallel"). A large number of small computers working together is called "massively parallel processing."

One of the complications of designing computers like this is that the small computers need to send information to one another. (For example, the results of the computation in one small computer may be needed by some of the others to complete their computations.) Wires and electronic circuits must be included to

Is This Useful?

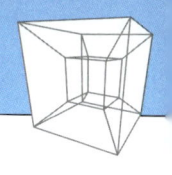

transfer data between the computers. This, of course, increases the cost of the computer.

The designers are confronted with the conflicting goals of minimizing the number of connections to keep the cost down, and adding more connections so the computers can exchange information more quickly. (When a small computer is waiting for information from another one, it isn't running its program and that computer time is wasted. Faster methods of exchanging information keep the small computers busier, which makes the massively parallel computer system more efficient.)

A low-cost design would be connecting all the computers to each other in a line, each computer passing the data along until it gets to the one that needs it. This would only require two connections for each computer. The problem is that it might take a long time for the data to move from the computer that has the data to the one that needs it.

The design with the ultimate performance would be to connect every computer to every other computer, but that would require each computer to have a large number of connections and there would be a huge pile

The Fourth Dimension and Beyond

of wires. This would be very expensive to build.

One computer scientist used the 12-dimensional cube to design the connections. In chapter 7 we learned that a 12D-cube has 4096 vertices and there are twelve lines at each vertex. In the system he designed, the computers were placed at the vertices of the 12D-cube and each computer had twelve connections. As with the low-cost solution, computers passed the data from one computer to the next, but it never had to pass through more than ten computers to reach its destination.

The computer system wasn't actually built in twelve dimensions, of course; the wires were connected between the computers in our 3D-world using the 12D-cube as a model. This clever use of hypercube math resulted in good performance at a reasonable cost.

Is This Useful?

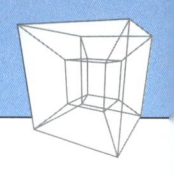

DATA ANALYSIS

Large amounts of data can be collected to describe our world. For example, a researcher studying plant growth may collect information such as this:

- temperatures during each day of the growing season
- amount of and temperature of rain
- strength of sunshine
- length of daylight for each day
- wind speed and direction
- type and amount of fertilizer added

Analyzing and understanding this much data is challenging.

One approach that mathematicians and scientists use is to think of each of the types of information as a dimension. You can then imagine creating a cube with enough dimensions for all of the data and storing the values in the cube. (We've learned how to this mathematically for any number of dimensions and computers make it easier to do.)

Here's a simple example. The following chart shows the heights of some plants; the heights are placed in

103

The Fourth Dimension and Beyond

the chart based on the amount of fertilizer added when the seeds were planted and the amount of water given to the plants each day.

Gallons of water per day

Pounds of fertilizer	0	10	20	30	40	50
0	0	2	4	5	3	1
2	0	3	6	8	7	5
4	0	5	8	12	15	14
6	0	4	7	10	12	10
8	0	3	6	8	8	6

Looking at this table, you probably read across the first row (fertilizer = 0) to see how the amount of water affected the plant growth. Then you might have looked at each of the other rows. You found that in each case, increasing the amount of water helped the plant grow larger up to a certain point. After that, there was apparently "too much water" for the plant.

When you picked out a single row to analyze, you "sliced" the data. In order to understand the information in the table, you look at one row, or slice, at a time. (You could also slice it vertically instead of horizontally to see how the amount of fertilizer affects plant growth for the various amounts of water.)

Is This Useful?

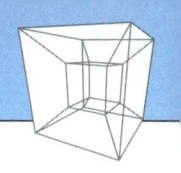

Pretend that the table is moved down through Flatland just as the square moved through Flatland. The Flatlander would "see" one row of the table at a time. The Flatlander would analyze one row—a slice—of the table at a time just as you did.

Mathematicians have studied how various shapes, with many different dimensions, are sliced as they pass through (or are "cut by") other shapes of various dimensions. They have then used these methods to help analyze complicated sets of data.

Beyond the Third Dimension by Thomas Banchoff includes a detailed example that shows techniques for visualizing data.

UNDERSTANDING OUR UNIVERSE

Astronomers and physicists are learning that more than three dimensions may be needed to explain our universe.

Scientists who study space believe that our universe is curved into the 4th dimension.

Imagine picking up Flatland and curving it. Do you think

The Fourth Dimension and Beyond

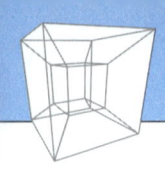

that the inhabitants of Flatland could determine that their world had been curved into the 3rd dimension?

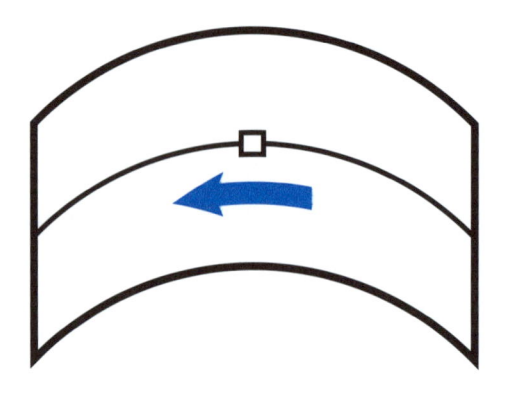

The square is following the line across Flatland. If you ask the square if it is going straight, it will say, "of course I am. I am following the line in a forward direction; I am not turning to the left or the right, so I am going straight."

The Flatlander, of course, doesn't realize that its world has been curved. It is going straight in its Flatland universe. But it is easily apparent to us that the Flatlander's universe is curved and, even though its path is "straight" in Flatland, it is not straight when we observe it in three dimensions.

If you continued to curve Flatland and rolled it into a

Is This Useful?

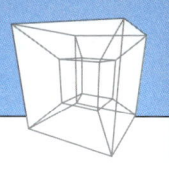

tube, the confused square would follow the "straight" path and eventually end up back where it started.

In a similar way, it is possible that our 3-dimensional universe is curved in the 4th dimension and when we think we are going straight, we are actually curving in the 4th dimension. In fact, if our universe is completely curved, we could send a rocket straight up into space and many centuries later, after traveling a curve in the 4th dimension, it would return to the earth.

THE HISTORY OF FLATLAND

Flatland: A Romance of Many Dimensions

Edwin Abbott Abbott wrote the original story about Flatland in 1884 and his book has been in print continuously since it was written.

Abbott lived in England in the late 1800s, a time when Victorian English society was based on class structures and women were assumed to be inferior to men. Abbott, a clergyman and schoolmaster, fought against these injustices. In Flatland, he uses geometry to create imaginary lands and vividly describes their inhabitants. His book is both a brilliant introduction to a mathematical concept and a satire about Victorian England.

The story of Flatland in chapter 2 of this book is based on Abbott's book; it is a shorter, simpler version that provides an introduction to the mathematical ideas in Flatland without the social commentary.

If you enjoyed this book, you may also find the original Flatland interesting.

The Fourth Dimension and Beyond

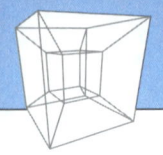

Since the introduction of Flatland, a variety of related works have been written. Some explain the Flatland book, some describe other imaginary worlds, and some explore the fourth dimension in more detail.

Visit www.BooksByJohnEikum.com for additional information about this exciting topic.

INDEX

A
Abbott, Edwin Abbott, 109–110
Analogies
    described, 31
    using, 32–37, 61–70, 86–88
Analyzing data, 103–105
Astronomy, 105–108

B
Banchoff, Thomas, 105
Base numbers, 75
Beyond the Third Dimension (Banchoff), 105
Boundaries, 53–56

C
Circles
    as 2-dimensional spheres, 67–68
    dimensions of, 16
    as parts of spheres, 24–25, 64, 69–70
Computers, 100–102
Containers, 52–53
Corners. *See* vertices
Cross sections of shapes, 36–37, 64

The Fourth Dimension and Beyond

Cubes
  0-dimensional, 58, 76
  1-dimensional, 58, 76
  2-dimensional, 58, 76
  3-dimensional, 54, 58, 76, 89–91
  4-dimensional, 49–52, 55, 57–58, 76, 91–96
  boundaries of, 54, 55
  parts of, 42–46, 75–79
  vertices of, 71–72

D

Data analysis, 103–105
Dimensions
  changes in shapes with increase in number of, 39, 59–60
  directions and, 7–9
  lack of, 29
  number of parts needed to make shapes in different, 57
  perpendicular, 12, 47–49, 81–84
  spatial, 97–98
  time as, 97–98
  using analogies to understand, 32–37, 61–70, 86–88
  See also specific dimensional shapes and worlds

Index

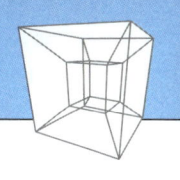

Directions
    perpendicular, 22
    types of, 8–10, 12–13, 49

E
Edges, described, 44
Einstein, Albert, 97, 98
Exponents, 75

F
Faces
    of 4-dimensional cubes, 56
    described, 43–44
    in different dimensional worlds, 53–56
Flatland. *See* 2-dimensional world
Flatland: A Romance of Many Directions (Abbott), 109–110
Flatlanders. *See* 2-dimensional shapes
Flipping of shapes, 64–66
Folding cubes
    3-dimensional, 89–91
    4-dimensional, 91–96
Forward/backward direction, 8–10
4-dimensional shapes
    in 3-dimensional world, 61–64, 91–96

The Fourth Dimension and Beyond

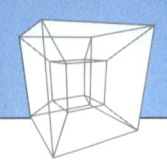

cubes, 55, 57–58, 76
described, 49–52
spheres as, 70
4-dimensional world
directions in, 47–49
perpendicular dimensions in, 83–84
4D-land. *See* 4-dimensional world
4D-landers. *See* 4-dimensional shapes
4D-up/4D-down, 49

H
Height, 15
Hidden lines, 45
Hypercube computers, 100–102
Hypercubes, 58

L
Left/right direction, 8–10
Length
of lines and points, 40
in 3-dimensional world, 15
Line segments
as boundaries, 54
boundaries of, 53
connecting, 71

Index

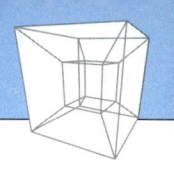

    as cubes, 58, 76
    in cubes, 43–44
    determining number meeting at each vertex, 72–74
    hidden, 45
    length and width of, 40
    perpendicular to each other, 11–12, 40–41
    vertices of, 72
Lineland. *See* 1-dimensional world
Linelanders. *See* 1-dimensional shapes
Lines, 40
    See also line segments

M
Massively parallel processing, 100–102
Mathematical notation, 75–79
Mutually perpendicular dimensions, 14, 15

N
Naming, consistency in, 58, 79

O
Objects. *See* shapes
1-dimensional shapes
    cubes, 58

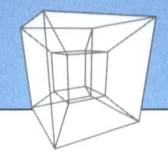

spheres, 66–67
1-dimensional world
    2-dimensional shapes in, 26–29, 32–37
    elements of, 59–60
    lines in, 40

P

Parallel processing, 100–102
Perpendicular dimensions
    in 2-dimensional world, 81–82
    in 3-dimensional world, 14–15, 82–83
    in 4-dimensional world, 83–84
    described, 47–49
    right angles and, 12
Perpendicular directions, 22
Perpendicular lines
    described, 11–12
    making shapes from, 40–41
Planes, described, 41
Pointland. *See* 0-dimensional world
Points
    in 0-dimensional world, 29, 39–40
    as boundaries, 53
    as cubes, 58, 76
    making spheres and, 66

Index

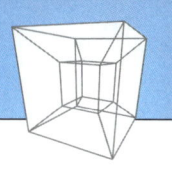

   size of, 29, 40
   as vertices, 73
Powers, 75

R
Radius, 66
Right angles, described, 10–11
Rotation of shapes, 64–66

S
Shapes
   being right next to, 32, 61
   as boundaries, 53–56
   cross sections of, 36–37, 64
   disappearance of, 36, 63, 94–96
   flipping over, 64–66
   increase in number of dimensions and, 39, 59–60
   number of parts needed to make, 57
   rotation of, 64–66
   seeing entire, 33–34, 61–62, 84–85
   seeing inside, 34–35, 62–63, 85–86
   See also specific shapes and shapes of specific dimensions
Sides, described, 41–42
Slicing data, 104–105

The Fourth Dimension and Beyond

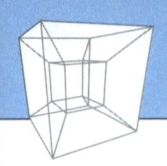

Spaceland. *See* 3-dimensional world
Spacelanders. *See* 3-dimensional shapes
Spacetime, 97–98
Spatial dimensions, 97–98
Spheres
    1-dimensional, 66–67
    2-dimensional, 67–68
    3-dimensional, 16, 68–70
    4-dimensional, 70
    circles and, 24–25, 64, 69–70
    dimensions of, 16
    making, 66–70
Squares
    in 2-dimensional world, 40–42
    in 3-dimensional world, 42–46
    as boundaries, 54
    boundaries of, 54
    as cubes, 58, 76
    vertices of, 71

T

Tesseracts, 57–58
3-dimensional shapes
    in 2-dimensional world, 20–25, 32–37, 61–62,
        63–64, 89–91, 94–95

Index

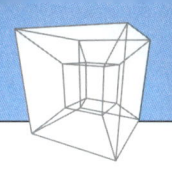

spheres, 16, 68–70
   in 3-dimensional world, 42–46
   vertices in, 44
3-dimensional world
   3-dimensional shapes in, 42–46
   4-dimensional shapes in, 61–64, 91–96
   directions in, 14–15, 16
   elements of, 60
   perpendicular dimensions in, 14–15, 82–83
Time as dimension, 97–98
2-dimensional shapes
   in 1-dimensional world, 26–29, 32–37
   in 2-dimensional world, 40–41
   circles, 16
   cubes as, 58
   spheres as, 67–68
   vertices in, 41–42
2-dimensional world
   2-dimensional shapes in, 40–41
   3-dimensional shapes in, 20–25, 32–37, 61–62, 63–64, 89–91, 94–95
   described, 17–19
   directions in, 10
   elements of, 59–60
   perpendicular dimensions in, 81–82

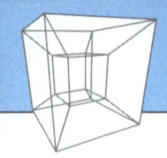

The Fourth Dimension and Beyond

U
Universe, understanding our, 105–108
Up/down direction, 12–13

V
Vertices
    in 2-dimensional shapes, 41–42
    in 3-dimensional shapes, 44
    of cubes, 71–72
    determining number of, 72–75
    points as, 73
    of squares, 71

W
Width
    of lines and points, 40
    in 3-dimensional world, 15

Z
0-dimensional shapes, 58
0-dimensional world
    elements of, 59–60
    points in, 29, 39–40